Heterocyclic Chemistry

Heterocyclic Chemistry

D. W. Young B.Sc. Ph.D.

School of Molecular Sciences, University of Sussex.

Longman
London and New York

LONGMAN GROUP LIMITED
London
Associated companies, branches and
representatives throughout the world
Published in the United States of America
by Longman Inc., New York

First published 1975

Library of Congress Cataloging in Publication Data
Young, Douglas W
Heterocyclic chemistry.

Includes bibliographies and index.
1. Heterocyclic compounds. I. Title.
QD400.Y68 547'.59 75-11739
ISBN 0-582-44253-2

Set in 10/12 pt. Times New Roman
and printed in England by
J. W. Arrowsmith Ltd., Bristol

Preface

This book began as a series of lectures to final year undergraduates at Sussex University. In Chapters 2 to 7 I have attempted to show how the interaction between heteroatoms and ring systems can give rise to the unique chemical properties of heterocyclic compounds and the subject matter is arranged with this aim in mind. Chapter 8 is an attempt to show how the chemical effects discussed in the early chapters can explain the action of biologically important compounds in nature and in medicine.

I would hope that the book will be useful to advanced undergraduate students of Chemistry and Biochemistry and I have assumed that the reader will have a basic knowledge of organic reactivity including some of the simpler aspects of orbital symmetry controlled concerted reactions and of resonance and m.o. concepts.

I would like to thank Dr Frank McCapra for reading the manuscript and for helpful discussion. Thanks are also due to my wife for her forbearance during the writing of the book and for typing the manuscript.

Contents

Reactions involving the heteroatom – Reactions typical of dienes – Electrophilic aromatic substitution reactions – Nucleophilic substitution and addition reactions – Reaction at the side chain of C-alkylated six-membered heteroaromatic compounds – Hydroxy and amino substituted six-membered heteroaromatic and related compounds – Other six-membered heteroaromatic compounds – Synthesis of six-membered heteroaromatic compounds – Key points – Further reading

Chapter 1

Introduction

Many organic compounds are heterocyclic, including naturally occurring compounds, pharmacologically active compounds and compounds of intrinsic theoretical interest. Heterocyclic compounds also play an important role in mediating many biological processes. It is, therefore, not surprising that much effort has been expended in studying their chemistry.

The genetic material DNA is heterocyclic, as are many useful alkaloids such as the anaesthetic cocaine, the narcotic nicotine, the antimalarial quinine and the amoebicide emetine. Useful synthetic heterocyclic compounds include the herbicide paraquat and the nylon feedstock caprolactam. The antibiotics penicillin and cephalosporin and a variety of vitamins, such as riboflavin and biotin, are heterocyclic. The structures and action of many of these compounds are discussed in Chapter 8.

Heterocyclic compounds are cyclic compounds with an element, or elements, other than carbon as part of the ring structure. By contrast, carbocyclic compounds are cyclic compounds in which only carbon atoms are involved in ring formation. Thus cyclohexane (**1**) is a carbocyclic compound and piperidine (**2**) is a heterocyclic compound.

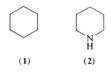

(**1**)   (**2**)

Heterocyclic compounds may be divided into two main types: aliphatic and aromatic. *Aliphatic heterocyclic compounds* are the cyclic analogues of amines, ethers, amides, enamines, etc., and have many properties in common with their acyclic analogues. In many aliphatic heterocyclic compounds, however, the fact that the heteroatom is constrained in a ring can have a considerable effect on the properties of the functional group and so aliphatic heterocyclic compounds may have unique properties. In Chapters 2 and 3, the chemistry of aliphatic heterocyclic compounds is looked at from the viewpoint of how the properties of heterocyclic amines, ethers, amides and of other functional groups may be affected by constraint in a ring.

Aromatic heterocyclic compounds are compounds which, having a heteroatom in a ring, have some of the properties which typify the chemistry of benzene. The heteroatom can play an extremely important role in determining the properties of compounds of this type, and in Chapters 4, 5, 6 and 7 the chemistry of this important class of heterocyclic compounds will be examined.

Nomenclature

Heterocyclic chemistry has been studied since the earliest days of chemistry and many compounds were named unsystematically. Since many heterocyclic ring systems were first observed in natural products, many of the names of these ring systems reflect the names of the plants in which compounds containing the ring systems were first found. These names passed into such general usage that it was impracticable to change them when systematic nomenclature was introduced; thus many of the more important ring systems are named non-systematically. Some examples of common heterocyclic compounds with non-systematic names are furan (**3**), pyrrole (**4**), thiophene (**5**), pyridine (**6**), indole (**7**) and quinoline (**8**). In general, when a non-systematic name is used in the following chapters, it will be accompanied by a structural diagram.

  (3)   (4)   (5)   (6)   (7)   (8)

In recent years the International Union of Pure and Applied Chemistry (IUPAC) has attempted to systematise heterocyclic nomenclature and all new ring systems, and many of the less common older heterocyclic systems, are now named systematically.

In the IUPAC system, the heteroatoms present in the compound are indicated by a prefix. Some of the prefixes used for the commoner heteroatoms are indicated in Table 1.1. When the suffix of the name begins with a vowel, then the vowel at the end of the heteroatom-defining prefix is omitted. The suffix in the name of the compound is used to denote ring size and, in some cases, degree of unsaturation, and common name endings are listed in Table 1.2.

Table 1.1

Prefixes

Heteroatom	O	N	S	Se	Te	P	As	Si
Prefix	Oxa	Aza	Thia	Selena	Tellura	Phospha	Arsa	Sila

Table 1.2

Suffixes

Number of atoms in ring	Suffix for fully unsaturated compound		Suffix for fully saturated compound	
	With nitrogen	With other heteroatom	With nitrogen	With other heteroatom
3	irine	irene	iridine	irane
4	ete	ete	etidine	etane
5	ole	ole	olidine	olane
6	ine	in		ane
7	epine	epin		epane

Having defined the rules, it is now possible to exemplify them by naming some heterocyclic compounds, as follows. The order of precedence of the commoner heteroatoms is oxygen > sulphur > nitrogen.

Aziridine   Oxirane   Thiirene   Azete   Oxetane 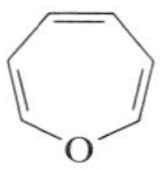   Oxepin

Thiazole    Isothiazole 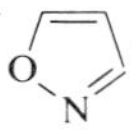   Isoxazole    Diazole

Numbering of heterocyclic compounds is necessary when substituents are present and the convention is to start numbering at the heteroatom and to number from it, round the ring, in such a way as to give the substituents the smallest numbers possible. Substituents are placed in alphabetical order. Thus the compound (**9**) is called 4-amino-2-ethylpyridine.

When there is more than one heteroatom present precedence for starting numbering is oxygen > sulphur > nitrogen, and the ring is numbered from the first heteroatom in such a way as to give the smallest possible number(s) to the other heteroatom(s). Hence the substituent plays no part in determining how the ring is numbered in such compounds, and compound (**10**) is named 4-methylisoxazole, compound (**11**) 5-methyloxazole and (**12**) 5-chloro-1,3-diazine.

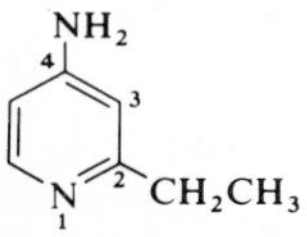

(**9**) 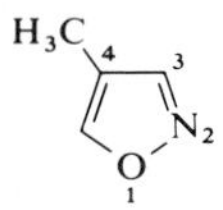   (**10**) 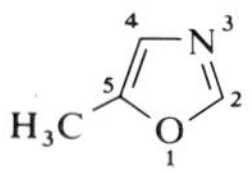   (**11**) 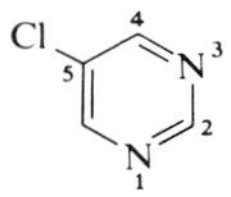   (**12**)

In most of the commoner and more straightforward cases, *fused polycyclic heterocyclic compounds* have non-systematic names. More complex examples are named in much the same way as fused-ring poly-carbocyclic compounds: by prefixing to the name of a component ring system, the names of other component parts of the molecule. Numbering of fused polycyclic heterocyclic compounds follows the numbering system devised for the corresponding carbocyclic system and so quinoline **(14)** and isoquinoline **(15)** follow the numbering system of naphthalene **(13)**; and indole **(17)** and isoindole **(18)** follow the numbering system of indene **(16)** even though, in some cases, this system leads to the single heteroatom not being accorded precedence.

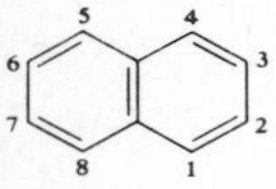

(13) 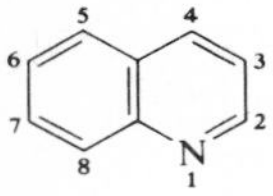   (14) 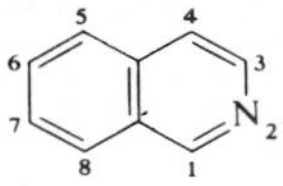   (15) 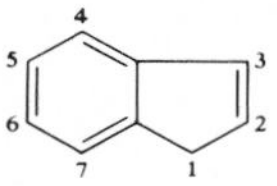   (16) 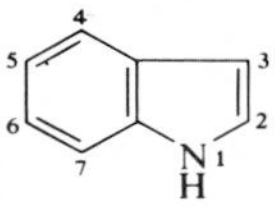   (17) 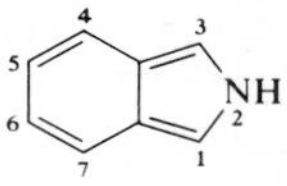   (18)

Chapter 2

Cyclic analogues of amines, ethers, thioethers and related compounds

Acyclic compounds such as amines and ethers have well-defined chemical and physical properties, but when the functional groups are constrained in a heterocyclic ring these properties may alter. Thus, although tetra-hydrofuran (**1**) and tetrahydropyran (**2**) have properties typical of acyclic ethers, ethylene oxide (**3**) has properties which differ considerably from those of its acyclic counterpart, dimethyl ether (**4**). It is the differences in properties of various functional groups caused by the groups being part of a ring system which are of special interest to the heterocyclic chemist.

$H_3C \quad CH_3$

| (1) | (2) | (3) | (4) |

§2.1 Bonding and geometry in cyclic compounds

§2.1.1 Three- and four-membered rings
In 1885 Baeyer, assuming rings to be planar, advanced his theory of ring strain. If a set of carbon atoms was constrained in a ring so that the bond angles varied from the "normal" tetrahedral value of 109·5°, then he considered that the resultant compound would have a strained ring and would be unstable and difficult to synthesise. We now know that cyclo-hexane and higher cycloalkanes have puckered non-planar rings and so do not have the strain predicted by Baeyer. The strain theory can, however, still be applied to three- and four-membered rings. Three-membered rings deviate most from the "normal" bond angle and so we would expect these to be the most reactive.

The classical theory of strain can explain many of the properties of small ring compounds, such as their participation in ring-opening reactions. Such reactions are accompanied by loss of ring strain and so a variety of reactions are made thermodynamically more likely in strained ring compounds than in the equivalent acyclic or larger ring analogues. In order to explain other more subtle properties of small-ring compounds, it is necessary to consider Molecular Orbital (M.O.) models.

5

The model developed by *Coulson and Moffitt* has been very useful in explaining the properties of small-ring compounds. By neglecting the principle of maximum overlap of orbitals, these workers calculated a model which is described pictorially in Fig. 2.1. The internal bond angles (θ) of 106° ($< 109.5°$) reflect more *p* character in the hybrid orbital than in an sp^3 hybrid orbital and this is compensated for by the external bond orbitals having less *p* character than sp^3. Table 2.1 summarises bond lengths and external bond angles of cyclopropane and some three-membered ring heterocyclic compounds and it is significant that the experimentally-derived external bond angles have values between those expected of sp^3-(109.5°) and sp^2-(120°) hybridised carbon atoms.

Fig 2.1

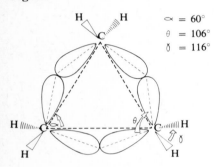

$\alpha = 60°$
$\theta = 106°$
$\delta = 116°$

Classical Baeyer strain theory predicts that four-membered ring compounds will be less strained than three-membered ring compounds. Three-membered rings must be planar, but with four-membered rings puckering can occur. Cyclobutane itself is puckered, as shown in (**5**) below. Puckering reduces the internal bond angles and so causes added ring strain but this is more than offset by the relief of non-bonded (steric) interactions between the eclipsed neighbouring hydrogen atoms which the planar model (**6**) would impose. Four-membered heterocyclic compounds such as oxetane (**7**) and thietane (**8**) have been shown to be planar and this is thought to be due to the reduction in the number of non-bonding interactions when a divalent heteroatom replaces a methylene group. Reintroduction of non-bonding interactions, such as occurs on oxidation of thietane (**8**) to the sulphone (**9**), causes the ring to become puckered once more.

	Steric interaction			
(**5**)	(**6**)	(**7**)	(**8**)	(**9**)

Table 2.1

Some molecular dimensions of three-membered rings

Compound	Bond lengths (Å)			External bond angle
	Cyclic Compound		Acyclic equivalent	
	C—C	C—X	C—X	H—C—H
Cyclopropane $$\begin{array}{c} CH_2 \\ H_2C-CH_2 \end{array}$$	1·53	—	1·54	116°
Ethylene oxide $$\begin{array}{c} O \\ H_2C-CH_2 \end{array}$$	1·47	1·44	1·42	116·4°
Aziridine $$\begin{array}{c} H \\ N \\ H_2C-CH_2 \end{array}$$	1·48	1·48	1·47	116·6°
Thiirane $$\begin{array}{c} S \\ H_2C-CH_2 \end{array}$$	1·49	1·82	1·81	116·0°

The Coulson–Moffitt treatment has been extended to cyclobutane and, although bonds are closer to sp^3 in character, there is still a small excess of p character in the ring bonds, compensated for by a diminution of p character in the external bonds. A pictorial representation of this model is shown in Fig. 2.2.

Fig 2.2

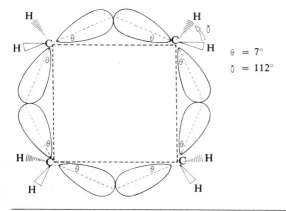

$$\theta = 7°$$
$$\delta = 112°$$

§ 2.1.2   *Five and six-membered rings*

There is little ring strain in rings of five or more atoms. Severe eclipsing of C—H bonds in cyclopentane makes a planar ring (Fig. 2.3 *a*) less likely than a puckered ring. Cyclopentanes are conformationally mobile but the envelope (*b*) and half-chair (*c*) forms predominate. Replacement of carbon by a heteroatom implies different bond angles and removal of C—H eclipsing interactions, and these can alter these preferred conformations.

Fig. 2.3

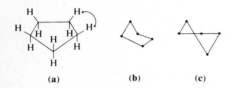

(a)   (b)   (c)

The cyclohexane ring system has been the subject of much conformational study and it is well recognised that the chair forms of Fig. 2.4*a* are preferred to the alternative boat conformation of Fig. 2.4*b*. There is a low but finite energy barrier to ring inversion between the two chair forms in Fig. 2.4*a*, and since 1,3-diaxial non-bonding interactions are fairly large, bulky substituents prefer equatorial (e) rather than axial (a) positions.

Fig. 2.4

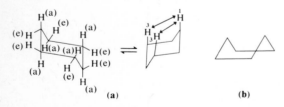

(a)   (b)

Although replacement of carbon by a heteroatom will cause changes in bond angles, six-membered heterocyclic compounds exist in chair conformations similar to those of cyclohexane. Intramolecular hydrogen bonding may determine the preferred conformation in substituted six-membered heterocyclic compounds (as shown in Fig. 2.5*a*) but, when there is no hydrogen bonding, substituents on carbon atoms not adjacent to the heteroatom adopt the expected equatorial conformation. Electronegative substituents on the carbon atom adjacent (α) to the heteroatom take up a conformation which is determined by the *anomeric effect*. This effect was first noted in carbohydrates where it was observed that there was an *axial* preference for acetoxyl, methoxyl and halogen groups if they were situated on the carbon adjacent to the pyranose ring oxygen (cf. Fig. 2.5*b*). The effect now appears to be general and is thought to be due, in part, to the fact that the dipole interaction of the ring oxygen and the electronegative group is more favourable when the electronegative group is axial (Fig. 2.5*b*) than when it is equatorial (Fig. 2.5*c*).

Fig. 2.5

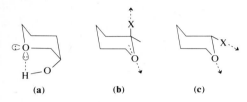

(a)   (b)   (c)

When nitrogen is the heteroatom, the ability of trivalent nitrogen to undergo pyramidal inversion (see § 2.4) allows piperidine rings to equilibrate by pyramidal inversion as well as by ring inversion as shown in Fig. 2.6. Pyramidal inversion is not possible with the more conformationally rigid phosphorus and arsenic analogues.

Fig. 2.6

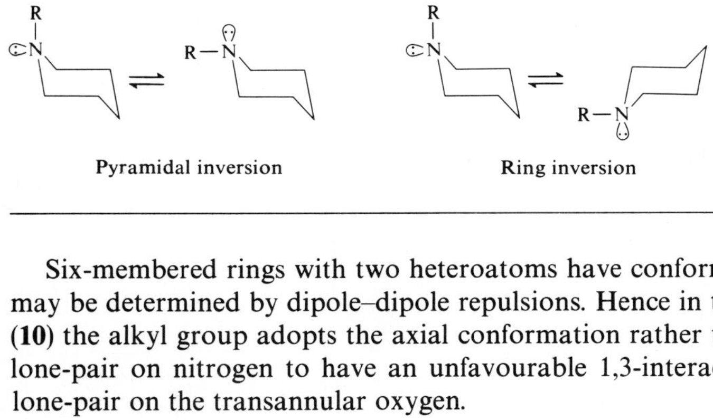

Pyramidal inversion   Ring inversion

Six-membered rings with two heteroatoms have conformations which may be determined by dipole–dipole repulsions. Hence in the compound (**10**) the alkyl group adopts the axial conformation rather than allow the lone-pair on nitrogen to have an unfavourable 1,3-interaction with the lone-pair on the transannular oxygen.

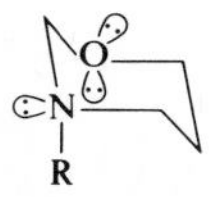

(**10**)

§ *2.1.3 Larger ring saturated heterocyclic compounds*
These have received attention recently with the publication of work on the chemistry of crown ethers. These are cyclic polyethers and if the ring

is sufficiently large, the heteroatoms may trap metal cations in the central cavity by ion–dipole interaction as shown in the rubidium salt (11) below. The outside of the structure consists of the hydrocarbon part of the ring and so the complex will be soluble in organic solvents. Crown ethers may, therefore, be used to make various metal salts soluble in organic solvents, and the solubilisation of $KMnO_4$ in benzene by crown ethers has been employed to increase the efficiency of permanganate oxidations.

(11)

The ratio of the size of the internal cavity of crown ethers to the size of the metal ion is important for complex stability and crown ethers can show a remarkable specificity for metal ions. Cyclic polyethers can discriminate between sodium and potassium ions and may be used in separation of these ions. Large ring antibiotics such as valinomycin are thought to affect ionic transport across cell membranes by acting as crown ethers in complexing the ions.

§ 2.1.4   Bridged-ring saturated heterocyclic compounds

These have very similar geometries to their carbocyclic counterparts. Certain unusual features such as the basicity of quinuclidine (see § 2.3) may, however, result from the geometry of the system. Large-ring bridged amines with nitrogen at the bridgehead are of especial interest as, if the rings are large enough, nitrogen inversion can occur and the lone-pairs may be either inside or outside the bridge structure (see Fig. 2.7a). Hydrochloride salts can stabilise the conformer with the hydrogen within the cage by encapsulating a chloride anion (see Fig. 2.7b).

Systems in which a second heteroatom is incorporated in the bridges give a caged crown ether molecule which can encapsulate metal ions within the cage. Such complexes have been termed "cryptates" and an X-ray study has shown a rubidium cryptate to have the geometry shown in Fig. 2.7c.

§ 2.2   Conjugative effects in cyclic compounds

Cyclopropane has some π-character in the plane of the ring due to the extra p contribution to the hybridisation of the ring bonds. The $\pi \rightarrow \pi^*$

Fig. 2.7

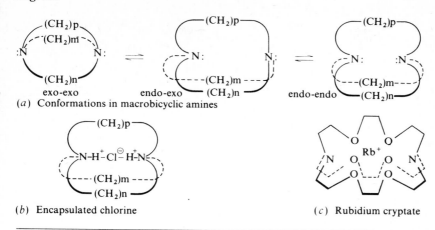

(a) Conformations in macrobicyclic amines

(b) Encapsulated chlorine

(c) Rubidium cryptate

transition in the ultra-violet spectra of olefins appears at a higher wave-length when a cyclopropane ring is "conjugated" with the olefin, and oxiranes and thiiranes exhibit similar effects to cyclopropane in this respect.

§2.3 Basicity in saturated heterocyclic compounds

It has been known for some time that cyclic amines form complexes with the Lewis acid trimethylboron and that complex stability and base strength vary with the size of the amine ring in the order:

4M ring > 5 > 6 > acyclic amines > 3M ring amine

The order for four-, five- and six-membered amines can be explained by the consideration of the degree of *F-strain* in these compounds. F-strain or frontal strain refers to the non-bonding interaction between the bulky groups surrounding the lone-pair, and the bulky acid. This will prevent complex formation. In acyclic tertiary and secondary amines, conformational mobility will allow the alkyl groups round the nitrogen to interact sterically with the bulky acid. When the groups are "tied back" by inclusion in a ring they will, however, be less able to interact in "front" of the lone-pair. Since in smaller rings the alkyl groups are "tied back" more, there is less steric inhibition to complex formation and salts will be more easily formed and be more stable when formed. The three-membered ring aziridine will have least F-strain and so, on this criterion, we might expect it to be the most basic rather than the least basic of the compounds. Its low base strength can be explained by consideration of the hybridisation of the lone-pair which, on the basis of the Coulson–Moffitt model, will be between sp^2 and sp^3. The greater contribution of the lower lying s component to the hybrid will bring the lone-pair closer to the nucleus and so it will be less available for bonding to an acid. The sp^2 hybridisation in pyridine (12) and in oximes (13) by the same token makes them much weaker bases than are aliphatic amines.

(12)   (13)

A study of the electron donor properties of cyclic ethers has shown them to be in the same order as the cyclic amines, although cyclic sulphides (5M > 6M > 4M > 3M) do not seem to follow the same pattern.

The F-strain argument which holds for a bulky acid such as trimethyl-boron may be difficult to extrapolate to protic acids, towards which the amines have the same order of base strengths as for bulkier acids. There is some evidence from nmr studies that the electron densities on the hetero-atoms may follow the same order as the reported basicities. The high basicity of the bicyclic base quinuclidine (14) can readily be explained on the basis of F-strain since the three alkyl groups are "tied back" behind the lone-pair.

(14)

§ 2.4 Pyramidal inversion in nitrogen compounds

Because asymmetrically substituted amines had never been resolved, it was at one time thought that trivalent nitrogen might be planar. Dipole moment and X-ray studies, however, showed nitrogen to have the pyramidal geometry as in (15) and the lack of resolution was attributed to easy racemisation by pyramidal inversion with a very low energy barrier. This can be visualised as interconversion of the sp^3 hybridised (15) to (17) via an sp^2 hybridised transition state (16).

(15)   (16)   (17)

On the basis of strain theory, (15) with 109·5° bond angles must open to (16) with 120° bond angles so that, if the nitrogen atom were part of an aziridine ring, this process would lead to an increase in ring strain. Thus we might expect the energy barrier to pyramidal inversion to be higher in asymmetrically substituted aziridines, and nmr studies have shown that indeed this is the case.

Since electronegative substituents on nitrogen make the nitrogen more electronegative so that the lone-pair will be closer to the nucleus, we might expect electronegative substituents to increase the energy barrier to inversion still further. The first pyramidal nitrogen compounds to be separated were diastereoisomeric N-chloroaziridines in 1968. Furthermore, asymmetric synthesis of an optically active oxaziridine with the ring nitrogen as the sole asymmetric atom has been achieved, as follows:

$$\underset{H_3C}{\overset{}{}}\!\!N\!\!=\!\!C\!\!\overset{Ph}{\underset{Ph}{}} \xrightarrow{(+)\text{ percamphoric acid}} \underset{H_3C}{\overset{O}{}}\!\!N\!\!\overset{Ph}{\underset{Ph}{}} \qquad (2.1)$$

In elements of the third and higher rows of the periodic table, hybridisation between s and p orbitals becomes more difficult and bonding tends to be more akin to pure p. Thus for phosphorus and higher elements in group five of the periodic table, the sp^2 transition state necessary for inversion and akin to (16) above is more difficult to achieve. Such elements are, therefore, configurationally more stable and less prone to inversion than is nitrogen. Bulky groups on the third valency of phosphorus tend to destabilise the ground state by non-bonding interactions and so inversion can be observed on heating the *tert*-butylphosphetane (18). There are reports that the strained oxonium and sulphonium salts, (19) and (20) respectively, have higher inversion barriers than the acyclic analogues.

(18)   (19)   (20)

§2.5 Reactions of saturated heterocyclic compounds as "normal" amines, ethers and thioethers

Reactions of saturated heterocyclic compounds with rings containing five or more atoms are fairly typical of the reactions of the acyclic analogues. Three- and four-membered ring saturated heterocyclic compounds are strained and can undergo quite different reactions from acyclic ethers, amines and thioethers to relieve this strain. There are, however, reactions, even with three-membered heterocyclic compounds, which might be defined as "normal". Aziridine will react as a nucleophile with a variety of electrophiles as follows:

$$\triangleright\!\!NH + RBr \longrightarrow \triangleright\!\!\underset{H^+}{\overset{+}{N}}\!\!-\!\!R \;\; Br^- \qquad (2.2)$$

$$\triangleright\!\!NH + H_2C\!\!=\!\!CHCN \longrightarrow \triangleright\!\!N\!\!\diagdown\!\!\diagup\!\!CN \qquad (2.3)$$

$$\triangleright\!\!NH + \triangle\!\!\overset{O}{} \longrightarrow \triangleright\!\!N\!\!\diagdown\!\!\diagup\!\!OH \qquad (2.4)$$

$$\triangleright\!\!NH \xrightarrow{RCH_2COCl \text{ or } RCH=C=O} \triangleright\!\!N\!\!\underset{\overset{\|}{O}}{\overset{}{}}\!CCH_2R \qquad (2.5)$$

  In alkylation and acylation reactions ((2.2) and (2.5) above), aziridine behaves in much the same way as does an acyclic amine. Michael addition as in (2.3) is also a "normal" reaction while, in reaction with epoxides (2.4), the more electronegative oxygen makes fission of the epoxide ring more likely and the aziridine acts as a nucleophile. Nitrosation of aziridine occurs "normally" at very low temperatures but at room temperature the N-nitroso derivative will break down as in reaction (2.6). Peralkylation of aziridine yields unstable quaternary salts which are prone to ring-opening reactions. The more stable four-membered azetidines form stable quaternary salts. Neither thiiranes nor thietanes can be alkylated without undergoing ring-opening reactions as in (2.7) and (2.8) below.

$$\triangleright NH \xrightarrow{\text{NOCl}/-60°C} \triangleright N-NO \xrightarrow{\text{Room temp.}} \begin{matrix} CH_2 \\ \| \\ CH_2 \end{matrix} + N_2O \tag{2.6}$$

$$\underset{\triangle}{\overset{S}{\triangle}} \xrightarrow{\text{CH}_3\text{Cl}} \underset{\triangle}{\overset{\overset{\displaystyle CH_3}{|}}{\overset{S^\oplus}{\triangle}}} Cl^\ominus \longrightarrow ClCH_2CH_2SCH_3 \tag{2.7}$$

$$\overset{S}{\boxed{}} \xrightarrow{\text{CH}_3\text{I}} \overset{\overset{\oplus}{S}-CH_3}{\boxed{}} I^\ominus \xrightarrow{\text{CH}_3\text{I}} ICH_2CH_2CH_2\overset{\oplus}{S}(CH_3)_2 \tag{2.8}$$

$$\overset{S}{\boxed{}} \xrightarrow{H_2O_2} \overset{S\rightarrow O}{\boxed{}} \xrightarrow{H_2O_2/\Delta} \overset{\overset{O}{\uparrow}}{\underset{(22)}{\overset{S\rightarrow O}{\boxed{}}}} \xrightarrow{\text{LiAlH}_4} \overset{S}{\boxed{}} \tag{2.9}$$
$$\qquad\quad (21)$$

  Thietanes can be oxidised to sulphoxides (21) and sulphones (22) as in (2.9) above. Acyclic thioethers undergo similar reactions and recently a thiirane has been oxidised in like manner. The sulphone (22) is unlike its acyclic counterpart in that it can be reduced to the parent thietane with lithium aluminium hydride. This is probably due to the relief of non-bonding interactions effected in this process (cf. § 2.1).

§ 2.6 Ring-opening reactions

Small-ring heterocyclic compounds have strained rings and relief of this strain makes ring-opening reactions likely. Calculations have estimated the strain energy of aziridine to be 58 Kjoule mole^{-1}; of oxirane 54 Kjoule mole^{-1}; and of thiirane 38 Kjoule mole^{-1}. These three-membered ring compounds not surprisingly undergo ring-opening reactions most readily. In ring-opening reactions, the most readily cleaved bond is that between carbon and the electronegative heteroatom, as indicated in (2.10) below. The positively-charged carbon produced in this process can then be attacked by nucleophilic reagents.

$$\underset{(CH_2)_n}{\overset{X}{\underset{\diagdown}{\overset{\diagup}{>}C}}} \longrightarrow \underset{(CH_2)_n}{\overset{\diagup}{>}C^\oplus} \diagdown X^\ominus \tag{2.10}$$

Three-membered ring heterocyclic compounds undergo ring-opening reactions with nucleophilic reagents and, in all of the cases studied, the attack has shown some degree of Walden inversion, the nucleophilic reagent attacking the carbon at the side opposite to the heteroatom "leaving group". This Walden inversion, although partial in some cases, implies an S_N2-type mechanism as depicted in (2.11) below. When asymmetrically-substituted heterocyclic compounds undergo such a reaction, steric hindrance causes attack of the nucleophile at the less substituted site as in (2.12). Such specificity is expected of S_N2-type processes where attack by the nucleophilic reagent is as important a process as bond-cleavage. Such a mechanism has been termed a *"push"* *mechanism*.

$$R_1 \overset{X}{\underset{R_2 \; Nu^{\ominus}}{\triangle}} R_3 R_4 \longrightarrow R_1 \overset{X^{\delta-}}{\underset{R_2 \; Nu^{\delta-}}{\triangle}} R_3 R_4 \longrightarrow R_1 \overset{X^{\ominus}}{\underset{R_2 \; Nu}{}} R_3 R_4 \tag{2.11}$$

$$CH_3\overset{O}{\overset{\frown}{CH}}{-}CH_2 \xrightarrow[\text{(2) } H_2O]{\text{(1) } PhMgBr} CH_3\overset{OH}{\underset{|}{CH}}{-}CH_2Ph \tag{2.12}$$

$$\underset{Ph}{\overset{Ph}{>}}\overset{O}{\triangle}CH_2 \longrightarrow \underset{Ph}{\overset{Ph^{\delta+}}{>}}\overset{O^{\delta-}}{\triangle}\overset{H}{\underset{H}{}} \xrightarrow{Nu^{\ominus}} \underset{Ph}{\overset{Ph}{>}}\underset{Nu}{\overset{}{}}{-}CH_2OH \tag{2.13}$$

In some cases bond-breaking may become more important than attack by nucleophile so that an S_N1-type, or *"pull"* *mechanism* may operate. Factors helping such a mechanism are the presence of carbonium-ion stabilising substituents and the use of ionic solvents, and asymmetrically substituted compounds will undergo ring-opening in a manner dictated by carbonium-ion stability rather than by steric effects, as in (2.13) above. Epoxides, with their more electronegative heteroatom, are more prone to ring-opening by a "pull" mechanism than are aziridines.

In *acid catalysed* ring-opening reactions, protonation of the basic heteroatom precedes ring cleavage. The heteroatom is made more electronegative by protonation and so a "pull" mechanism is much more likely to operate in acid-catalysed ring-cleavage reactions. It is not surprising, therefore, that base-catalysed reactions may yield quite different products from their acid-catalysed counterparts. In reaction (2.14) the strong nucleophilic reagent attacks at the less sterically hindered site in a "push" process, while in reaction (2.15) the better leaving group and poorer nucleophile make for a "pull" mechanism and attack at the alternative site is preferred.

$$PhCH\overset{O}{\overset{\frown}{}}CH_2 \xrightarrow{CH_3ONa} Ph\overset{OH}{\underset{|}{CH}}CH_2OCH_3 \; (70\%) \tag{2.14}$$

$$PhCH\overset{O}{\overset{\frown}{}}CH_2 \xrightarrow{H^+/CH_3OH} Ph\overset{OCH_3}{\underset{|}{CH}}{-}CH_2OH \; (90\%) \tag{2.15}$$

Four-membered ring heterocyclic compounds are less strained than their three-membered ring counterparts and so ring cleavage is less likely.

This is to say that there will be less contribution from the leaving group in ring cleavage reactions which will, therefore, require strongly nucleophilic reagents and occur by an S_N2 or "push" mechanism. Substitution is, therefore, normally at the less substituted of the carbon atoms attached to the heteroatom as in (2.16) and (2.17) below.

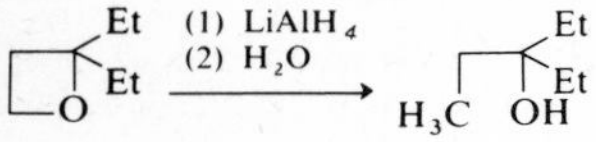

$$\text{(2.16)}$$

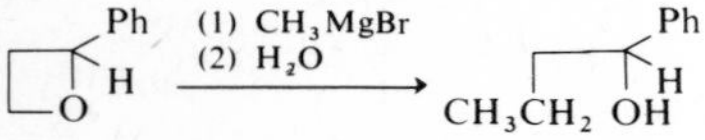

$$\text{(2.17)}$$

Acid-catalysed ring-opening reactions do occur readily with four-membered heterocyclic compounds. Normally the direction of attack is again at the least substituted carbon atom as in (2.18) below, but carbonium-ion stabilising groups such as phenyl can override this effect and the expected product of a "pull" mechanism may result, as in (2.19):

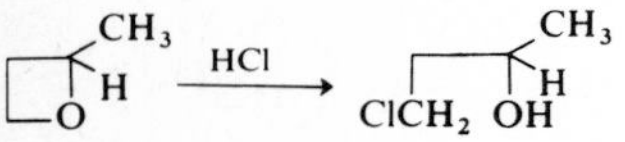

$$\text{(2.18)}$$

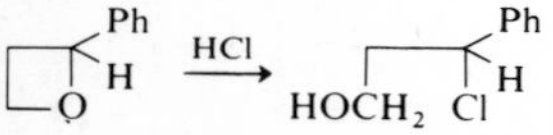

$$\text{(2.19)}$$

Ring-opening reactions include some reactions in which the heteroatom may be extruded, and olefins may be obtained from aziridines, oxiranes, thiiranes and some sulphoxides, as follows:

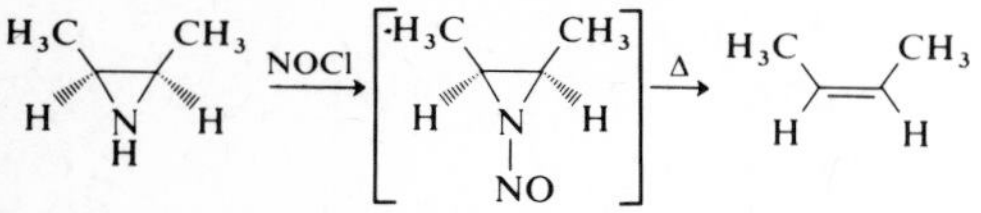

$$\text{(2.20)}$$

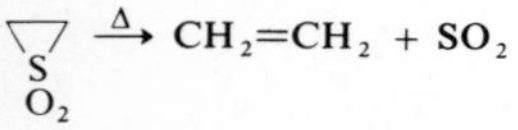

$$\xrightarrow{\Delta} CH_2=CH_2 + SO_2 \qquad \text{(2.21)}$$

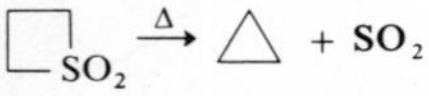

$$\xrightarrow{\Delta} \triangle + SO_2 \qquad \text{(2.22)}$$

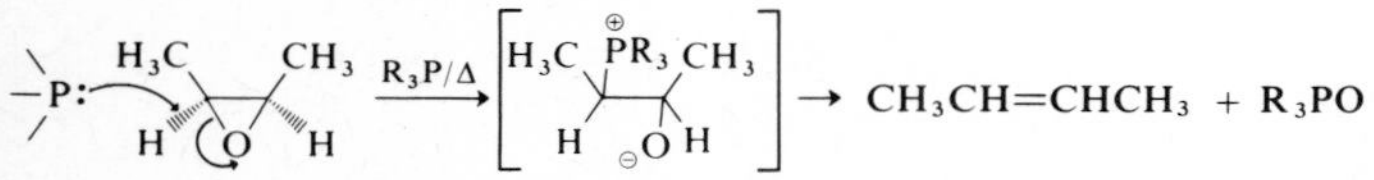

$$\xrightarrow{R_3P/\Delta} \; CH_3CH=CHCH_3 + R_3PO \qquad \text{(2.23)}$$

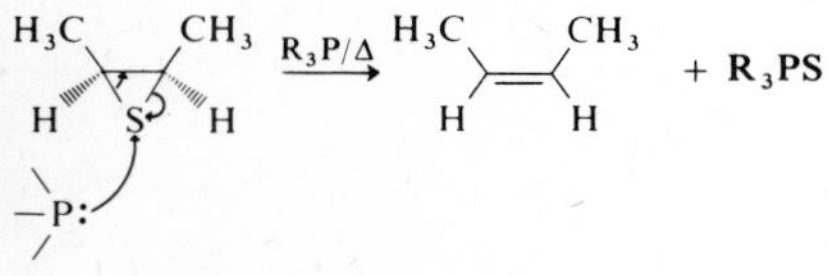

$$\xrightarrow{R_3P/\Delta} \; + R_3PS \qquad \text{(2.24)}$$

In the ring-opening reactions which we have so far considered, the bond between carbon and the heteroatom is cleaved. In some concerted reactions, however, orbital symmetry considerations may result in carbon–carbon bonds being cleaved. In reaction (2.26), the ring system (23) may be considered analogous to cyclobutene and so an electrocyclic ring-opening analogous to reaction (2.25) may occur. Since these reactions involve 4π-electrons, they will be thermally conrotatory as indicated in (2.25). When the aziridine (24) was pyrolysed and the 1,3-dipolar ring-cleavage product (25) was trapped by a dienophile as in (2.27), then the stereochemistry of the product was found to be that expected of a conrotatory ring-cleavage. Epoxides will undergo similar reactions, as shown in reaction (2.28).

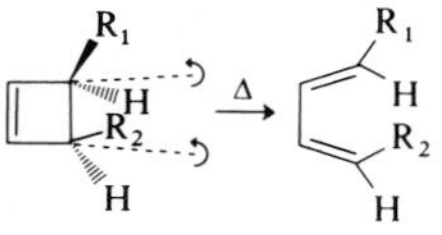

$$(2.25)$$

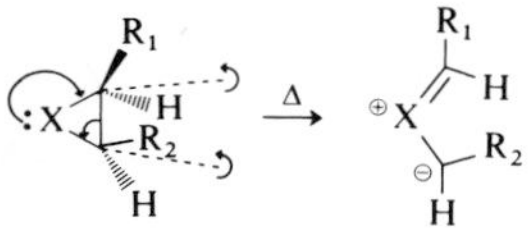

$$(2.26)$$

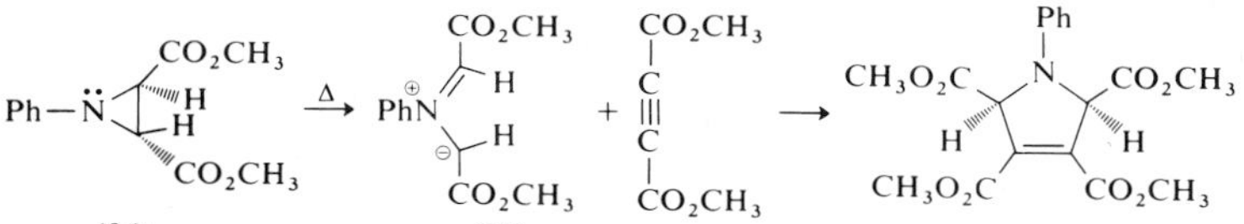

(23)

$$(2.27)$$

(24)   (25)

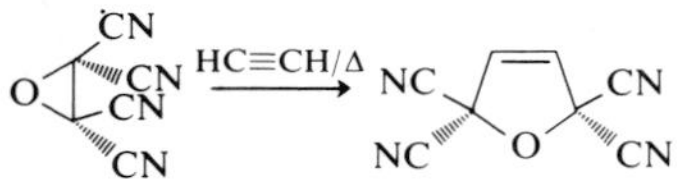

$$(2.28)$$

§ 2.7   Ring-opening reactions accompanied by rearrangement

Heterocyclic compounds undergo a variety of rearrangements which are completely analogous to rearrangements of the carbocyclic analogues. Thus the well-known vinylcyclopropane to cyclopentene rearrangement (2.29) has the heterocyclic analogues (2.30), (2.31) and (2.32). These have been used in the synthesis of five-membered heterocyclic compounds. The reaction has also been used with the azetidine (26), but since there is less relief of strain involved more severe conditions are required to achieve the reaction (2.33). Other rearrangements such as (2.34) have analogies in carbocyclic chemistry.

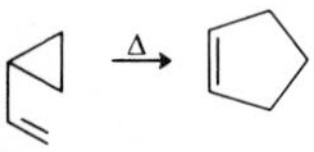

$$(2.29)$$

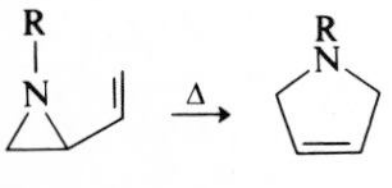

$$(2.30)$$

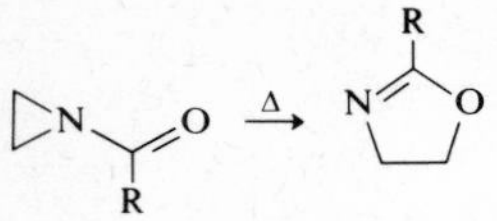

(2.31)

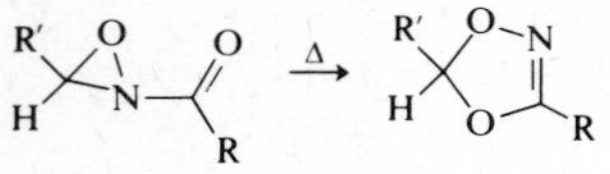

(2.32)

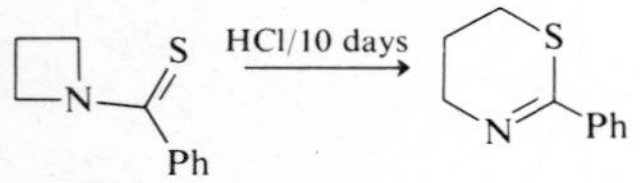

(2.33)

(26)

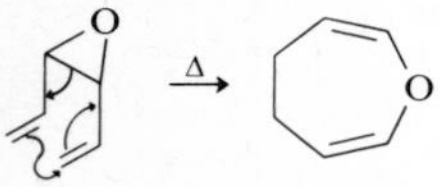

(2.34)

The vinylcyclopropane to cyclopentene rearrangement may be initiated by nucleophilic attack and reactions such as (2.35) and (2.36) have been used synthetically.

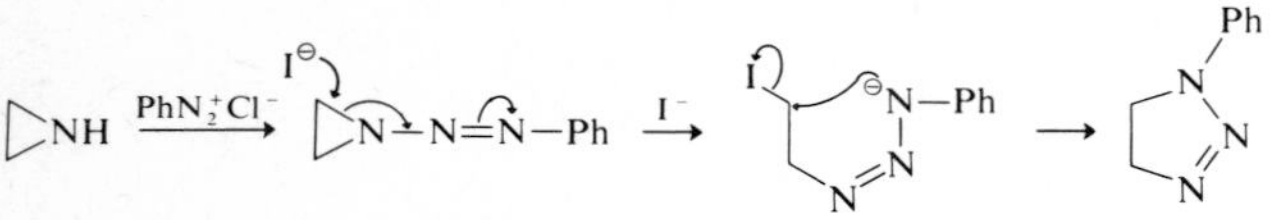

(2.35)

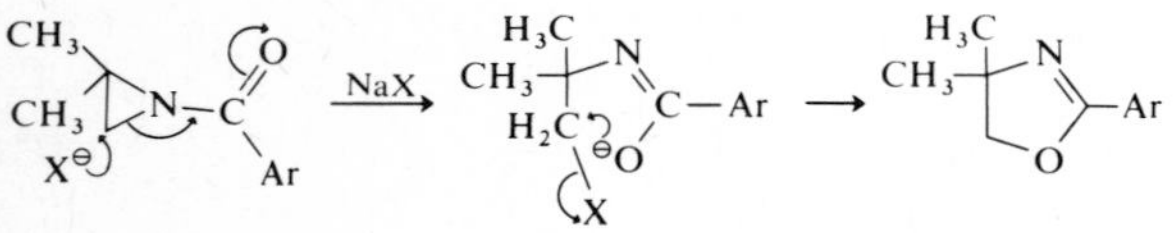

(2.36)

Oxiranes open very readily in acid to yield intermediate carbonium ions. These undergo the many rearrangements of carbonium ions and there are many instances of rearrangements of epoxides. The following are but two examples.

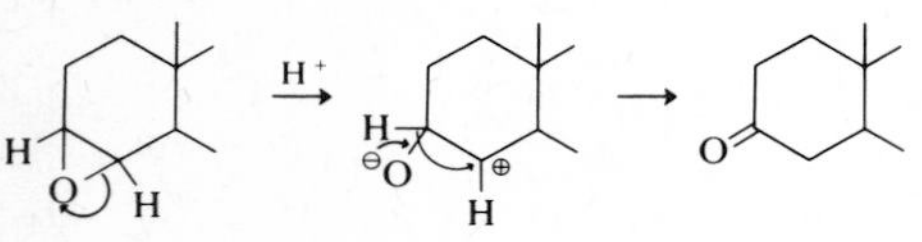

(2.37)

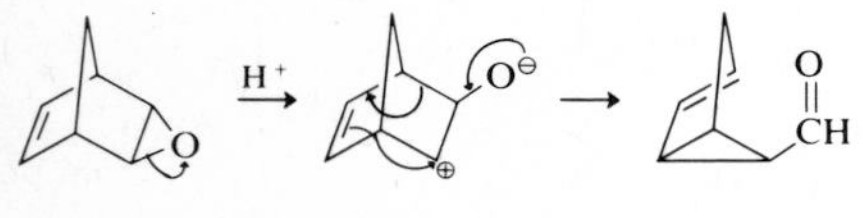

(2.38)

§ 2.8   Synthesis of saturated heterocyclic compounds

Saturated heterocyclic compounds may be synthesised in a variety of ways. These may be divided into two broad categories: intramolecular cyclisation and cycloaddition reactions.

(i) *Intramolecular cyclisation reactions* are useful methods of effecting synthesis of heterocyclic compounds. Often they are simply intramolecular versions of common methods of synthesis of their acyclic counterparts. The Williamson ether synthesis (2.39) is a useful method of synthesising acyclic ethers and cyclic ethers may be made in the same way by intramolecular Williamson ether synthesis using a halohydrin. Thus oxiranes (2.40) and oxetanes (2.41) may be made as shown. Intramolecular alkylation of amines will similarly yield aziridines (2.42), azetidines (2.43) and other cyclic amines.

$$ROH + R'Br \xrightarrow{\text{Base}} ROR' \qquad\qquad (2.39)$$

$$\begin{array}{cc} CH_2-CH_2 \\ | \quad\;\; | \\ OH \quad Br \end{array} \xrightarrow{\text{Base}} \triangledown_O \qquad\qquad (2.40)$$

$$\begin{array}{c} CH_2-CH_2Br \\ | \\ CH_2-O-H \end{array} \xrightarrow{\text{Base}} \underset{O}{\square} \qquad\qquad (2.41)$$

$$\begin{array}{cc} CH_2-CH_2 \\ | \quad\;\; | \\ Cl \quad\; NH_2 \end{array} \xrightarrow{\text{Base}} \underset{\underset{H}{N}}{\triangledown} \qquad\qquad (2.42)$$

$$\begin{array}{cc} CH_2-CH_2 \\ | \quad\;\; | \\ CH_2 \quad NH_2 \\ | \\ Br \end{array} \xrightarrow{\text{Base}} \underset{NH}{\square} \qquad\qquad (2.43)$$

These intramolecular reactions are subject to the restrictions imposed on cyclisations which give carbocyclic compounds. Activation energy for ring closure depends, in the main, on two factors. The first of these is the ring strain in the product of cyclisation. The greater this strain, the more difficult will be the cyclisation. The second factor is the probability of having the ends of the ring-forming chain in close proximity. As chain-length increases, so does the total number of conformations which will *not* allow ring closure and so ring forming ability falls off with chain length.

The effects of these two factors make three-membered rings relatively easy to form since, although the ring strain in three-membered rings is high, the probability of the ends of the chain being in the right conformation for ring closure is also high. Five-membered rings are also easy to form since the strain factor is minimal. Synthesis of four-membered rings is, however, very difficult since both strain and conformational factors are optimised for such cyclisations. Higher rings are harder to form due both to conformational factors and to non-bonding repulsions in these rings. That the three-membered ring is preferred to a possible four-membered ring is seen in the very clean reaction (2.44) where no azetidine product was observed. A study has been made of first order rate constants for the cyclisation of bromoalkylamines. This is summarised in Table 2.2.

$$\begin{array}{cc} CH_2-CH-Br \\ | \quad\qquad | \\ PhSO_2NH \quad CH_2-Br \end{array} \rightarrow \begin{array}{c} CH_2Br \\ \triangledown\!\!\!\diagup\, H \\ \underset{\underset{SO_2Ph}{N}}{} \end{array} \qquad\qquad (2.44)$$

Table 2.2

Reaction	Relative rate
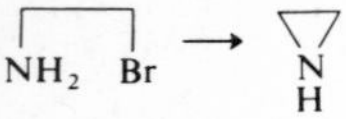 NH₂ Br → (aziridine, NH)	70
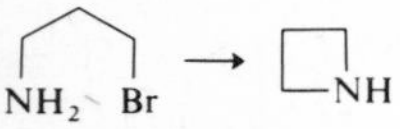 NH₂ Br → (azetidine, NH)	1
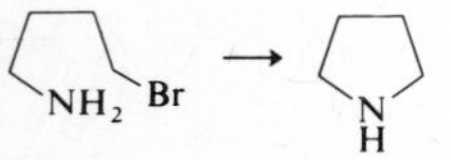 NH₂ Br → (pyrrolidine, NH)	60 000
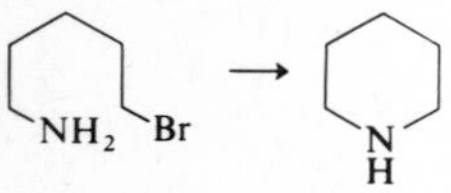 NH₂ Br → (piperidine, NH)	1000
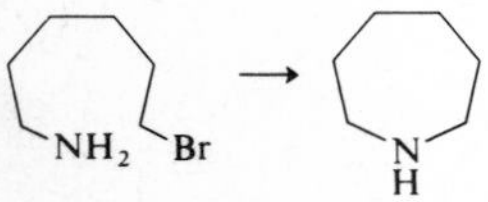 NH₂ Br → (azepane, NH)	2

Intramolecular cyclisation reactions of the general type (2.45) have been used in the synthesis of a great many saturated compounds as shown in reactions (2.46), (2.47), (2.48), (2.49) and (2.50), the intermediate (**27**) being generated *in situ* in many cases.

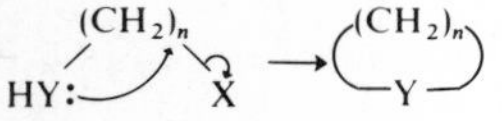

$$HY:\!\!-\!\!\overset{(CH_2)_n}{\diagup}\!\!-\!\!X \longrightarrow \overset{(CH_2)_n}{\diagdown}\!\!-\!\!Y\!\!-\!\!$$ (2.45)

(**27**)

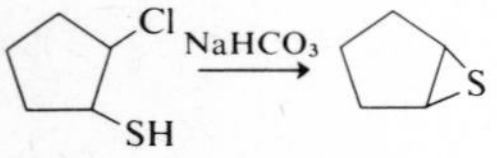

(2.46)

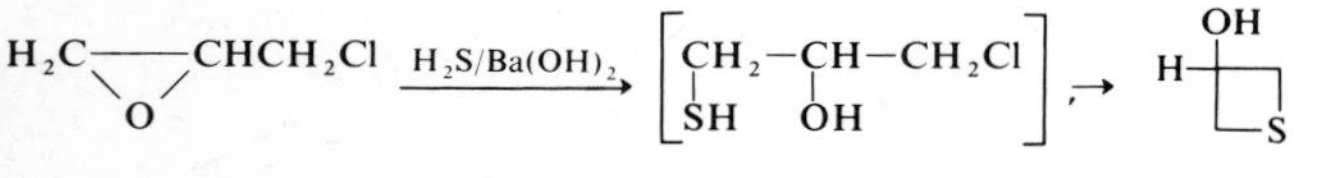

(2.47)

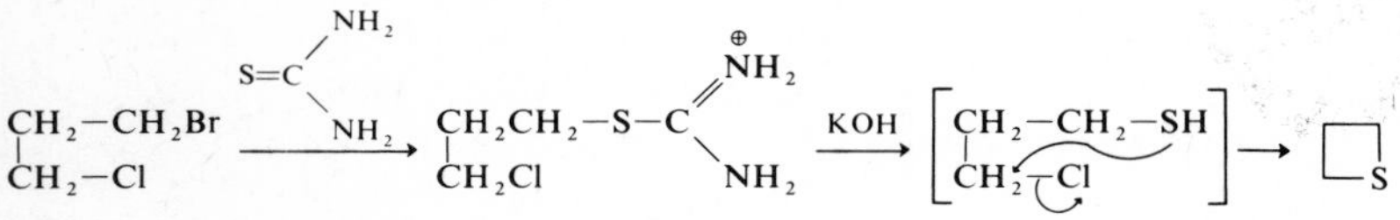

(2.48)

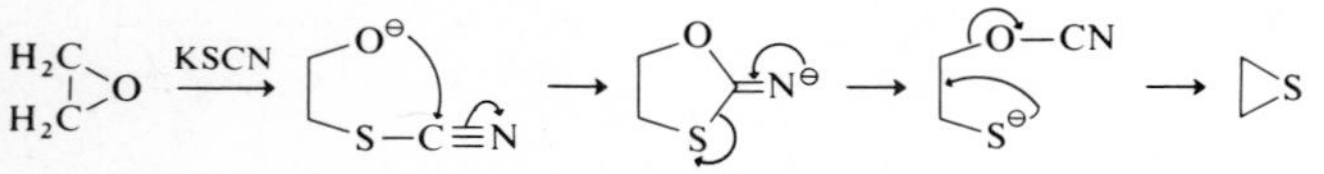

(2.49)

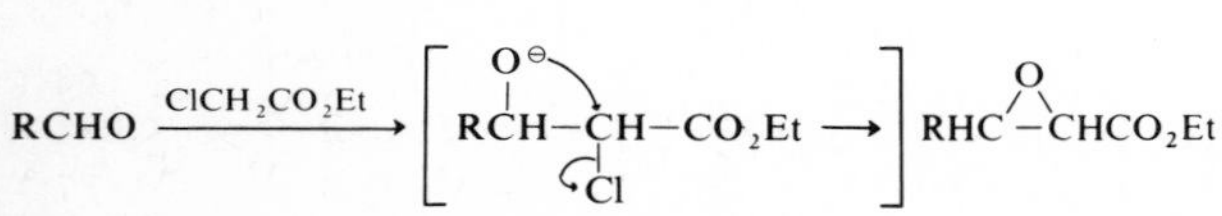

$$RCHO \xrightarrow{ClCH_2CO_2Et} \left[R\overset{O^{\ominus}}{\underset{\underset{Cl}{|}}{C}H\!-\!\overset{|}{C}H\!-\!CO_2Et} \right] \longrightarrow RHC\overset{O}{\diagup}CHCO_2Et$$ (2.50)

(ii) *Cycloaddition reactions* are useful in synthesis of carbocyclic and heterocyclic compounds, and a variety of compounds can be made in this way.

Three-membered heterocyclic compounds can be made by cycloaddition of a "one-atom" moiety to a "two-atom" moiety. The "one-atom" unit may be carbene or nitrene or may come from a peracid, and the "two-atom" unit may be imine, ketone or olefin. The reaction is normally stereospecific, the stereochemistry of the two-carbon unit being maintained in the cyclic product. Some examples where this type of synthesis is used to prepare epoxides, thiiranes, aziridines and other three-membered heterocyclic compounds are shown below ((2.51)–(2.56)). Two mechanisms are possible for reactions (2.53)–(2.56). In one mechanism the diazo compound or the azide can lose nitrogen to form a diradical species known as a carbene ($R_2C:$) or a nitrene ($R—N:$) respectively. These species will add to the olefin in a cycloaddition reaction. In the alternative mechanism the diazo compound or azide will react directly with the olefin in a 1,3-dipolar addition reaction such as (2.62) or (2.63) on pages 22 and 23. This reaction will yield a five-membered heterocyclic compound which will lose nitrogen on heating (cf. (2.62) and (2.63)) to yield the three-membered heterocyclic compound.

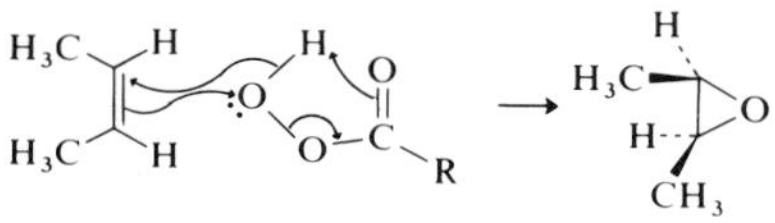

$$\text{(structure)} \longrightarrow \text{(epoxide with } H_3C, H, CH_3, O) \tag{2.51}$$

$$R\text{-CH=N-}{}^tBu + \text{(peracid)} \longrightarrow \text{(oxaziridine, } N\text{-}{}^tBu) \tag{2.52}$$

$$\underset{O}{\overset{R_1}{\underset{}{}}}\!\!C\!\!\underset{}{\overset{R_2}{}} + CH_2N_2 \xrightarrow{\Delta} \text{(epoxide, } R_1, R_2, O, CH_2) \tag{2.53}$$

$$\underset{S}{\overset{Ph}{\underset{}{}}}\!\!C\!\!\underset{}{\overset{Ph}{}} + Ph_2CN_2 \xrightarrow{\Delta} \text{(thiirane, Ph, Ph, Ph, Ph, S)} \tag{2.54}$$

$$\underset{R_3}{\overset{R_1}{\underset{}{}}}\!\!C\!\!=\!N\!\!\underset{}{\overset{R_2}{}} + CH_2N_2 \xrightarrow{\Delta} \text{(aziridine, } R_2, R_1, CH_2, N\text{-}R_3) \tag{2.55}$$

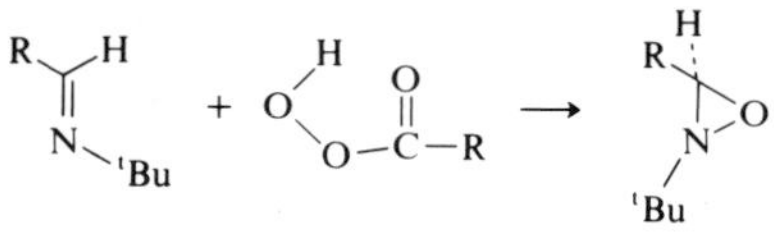

$$\underset{R_2,\,H}{\overset{R_1,\,H}{\underset{}{}}}\!\!C\!\!=\!C + ArN_3 \xrightarrow{\Delta} \text{(aziridine, } H, R_1, H, R_2, NAr) \tag{2.56}$$

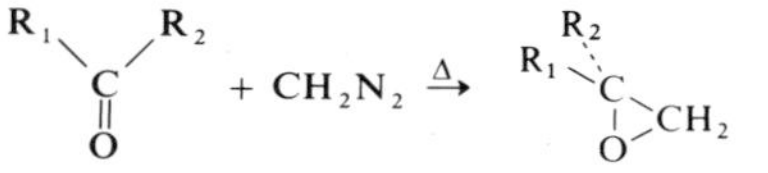

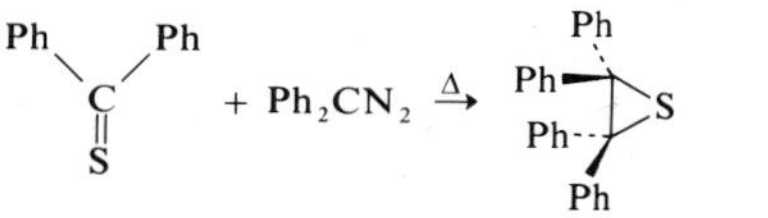

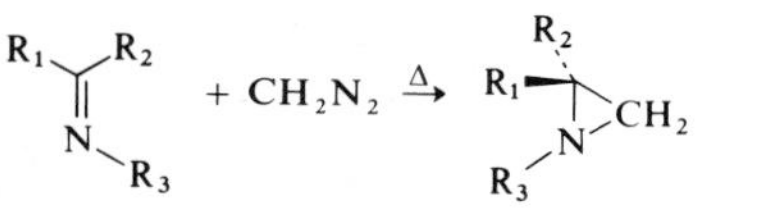

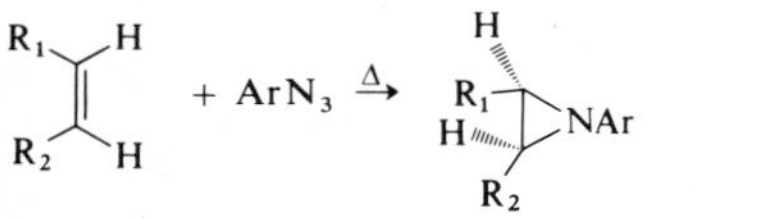

Peracid oxidation of olefins as in (2.51) is most readily achieved with electron-rich olefins such as tetrasubstituted olefins. Olefins substituted with electron-withdrawing groups such as enones will not react in this way but the oxiranes of enones may be readily synthesised by reaction with alkaline hydrogen peroxide as in (2.57) below:

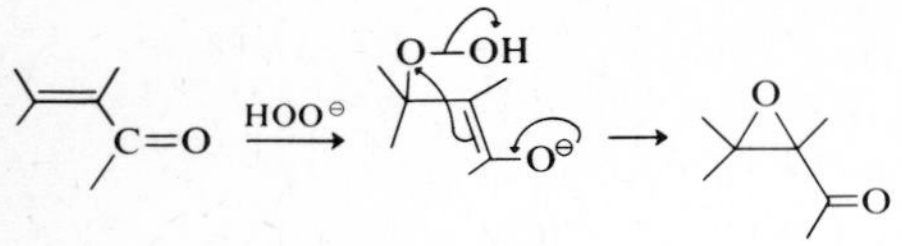

(2.57)

Four-membered heterocyclic compounds may be synthesised by cyclo-addition of two "two-atom" moieties. These reactions might be expected to be photochemical on the basis of orbital symmetry arguments although thermal processes are known and have been rationalised.

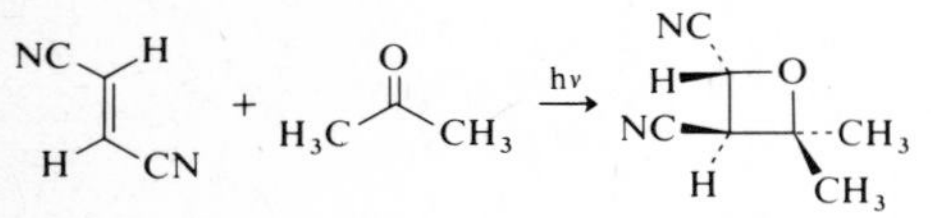

(2.58)

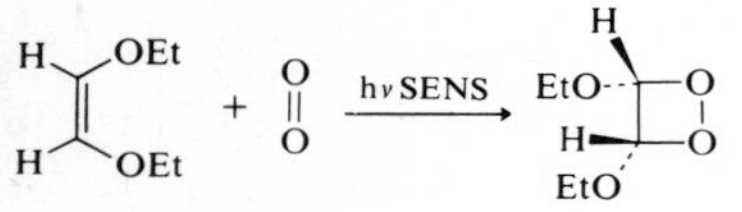

(2.59)

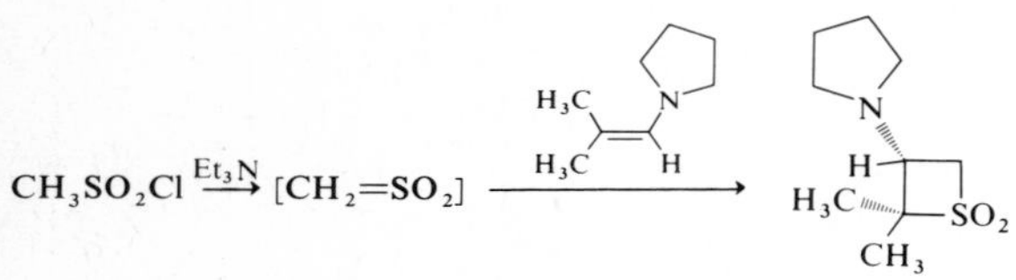

(2.60)

Five-membered heterocyclic compounds may be made by cyclisation of a "three-atom" moiety with a "two-atom" moiety and one of the most prolific sources of five-membered heterocyclic compounds is the 1,3-dipolar addition reaction. Reactions (2.27) and (2.28) in § 2.6 are examples of this type of reaction and five-membered nitrogen- and oxygen-containing heterocyclic compounds may be formed from aziridines and oxiranes respectively in these reactions. The reactions are expected to be thermally allowed on the basis of orbital symmetry arguments and further examples are (2.61), (2.62) and (2.63) below.

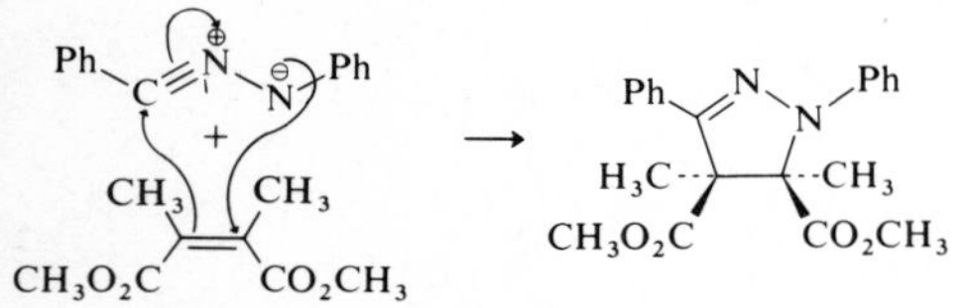

(2.61)

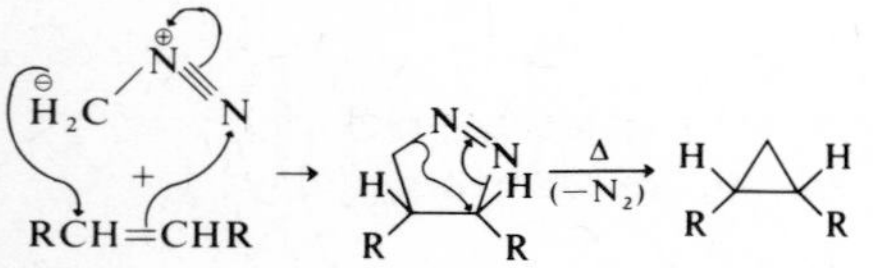

(2.62)

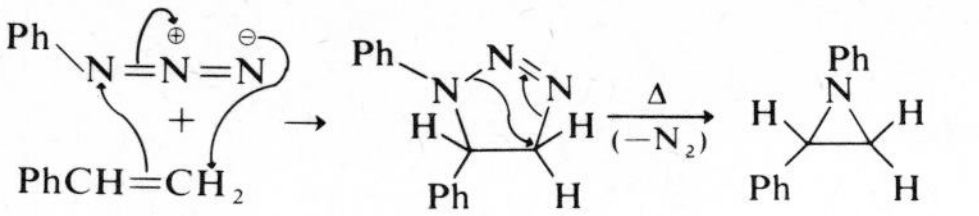

(2.63)

Cycloaddition of a diene to a "one-carbon" moiety may also be used to synthesise a five-membered heterocyclic compound.

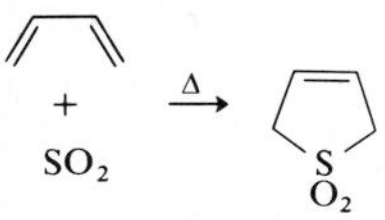

(2.64)

Six-membered heterocyclic compounds have been synthesised by the Diels–Alder reaction, a 4 + 2 cycloaddition, as in (2.65) or by addition of two "three-atom" units as in (2.66).

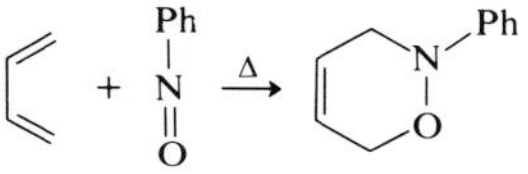

(2.65)

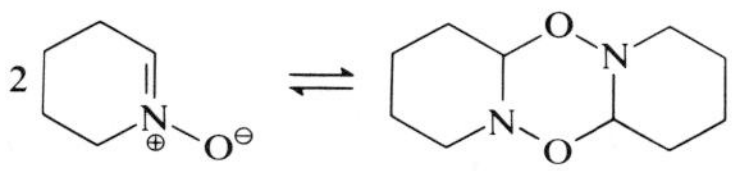

(2.66)

Key points

(1) The constraint of amines, ethers, thioethers, etc. in rings can profoundly affect their chemical and physical properties.

(2) Such property differences are most marked in small-ring heterocyclic compounds where ring strain, F-strain and hybridisation effects are maximised.

(3) Medium ring heterocyclic compounds can exhibit interesting conformational effects but their chemistry and that of most larger ring heterocyclic compounds is very much the same as that of their acyclic counterparts.

(4) Some larger ring heterocyclic compounds such as crown ethers exhibit very useful properties which result from the functional groups being part of a cyclic system.

(5) Ring opening reactions of small-ring heterocyclic compounds are extremely good examples of the requirements of nucleophile, site of attack and leaving group in nucleophilic substitution reactions.

(6) Syntheses of saturated heterocyclic compounds are of two types: cycloaddition reactions and intramolecular cyclisation reactions. The latter type of synthesis is very dependent on the chain length of the acyclic precursor.

Further reading

Conformational effects
E. L. Eliel, *Accounts of Chemical Research*, 1970, **3**, 1.

F. G. Riddell, *Quarterly Reviews*, 1967, **21**, 364.

Nitrogen inversion
J. M. Lehn, *Fortschritte der Chemischen Forschung*, 1970, **15**, 312 (in English).

Crown ethers
C. J. Pedersen and H. K. Frensdorff, *Angewandte Chemie International Edition*, 1972, **11**, 16.

Orbital symmetry
J. J. Vollmer and K. L. Servis, *J. Chem. Ed.*, 1968, **45**, 214.

General chemistry
Heterocyclic Compounds with Three- and Four-Membered Rings, (ed. A. Weissberger) Interscience, New York, 1964.
Heterocyclic Compounds, vol. 1 (ed. R. C. Elderfield) Wiley, New York, 1950.

Chapter 3

Unsaturated aliphatic heterocyclic compounds

Introduction of unsaturation to small ring compounds with the require-
ment of a 120° bond angle for sp^2 hybridised atoms will inevitably increase
ring strain. Unsaturated small-ring heterocyclic compounds will, therefore,
be more unstable than their saturated counterparts and will, in conse-
quence, be more difficult to prepare. The synthesis and reactions of
small-ring unstable compounds has been a popular field of research in
recent years.

§ 3.1   Olefinic heterocyclic compounds

Many three- and four-membered imines, enamines, enol ethers and
thioenol ethers have been prepared. Some of these are of theoretical
interest since they are iso-π-electronic with the cyclopropenyl carbonium
ion (**1**), cyclobutadiene (**2**) or the cyclopentadienyl anion (**3**) and so are
potentially aromatic or antiaromatic (see § 4.1 for an explanation of these
terms). The compounds will have considerable ring strain and so, considerations of aromaticity apart, their properties should be atypical of
either their acyclic or their cyclic but saturated counterparts.

(1)   (2) 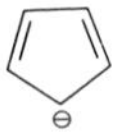   (3)

§ 3.1.1   Azirines
Two possible azirine structures can be written, the imine, 1-azirine (**4**),
and the enamine, 2-azirine (**5**):

(4) 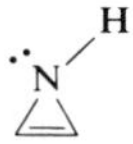   (5)

2-Azirine will be isoelectronic with cyclobutadiene, having four
π-electrons in planar cyclic conjugation, and so it will be resonance
destabilised or antiaromatic. This instability, coupled with the considerable ring strain, makes the synthesis of 2-azirines very unlikely and,

although there have been reports of the preparation of 2-azirines, none has stood up to critical study. There is, however, evidence that the 2-azirine (**7**) is an intermediate in the pyrolysis of the triazolines (**6**) and (**8**), where both compounds yield identical mixtures of products which include the 1-azirines (**9**) and (**10**).

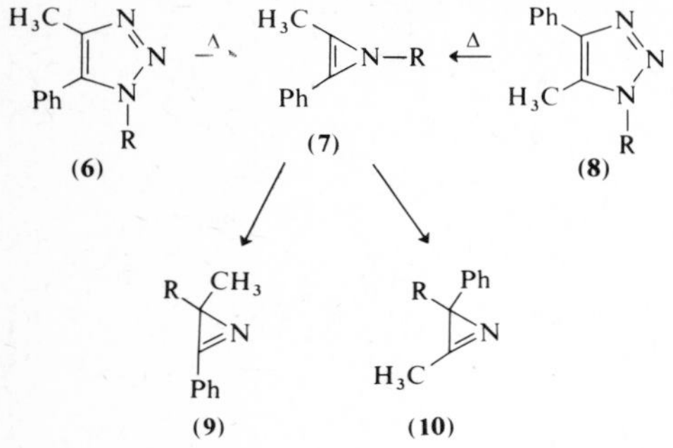

(3.1)

1-Azirines should be very strained but they are not antiaromatic and therefore should be much more likely to be synthesised than 2-azirines. In 1932 Neber postulated a 1-azirine structure for an intermediate which he had isolated in the base-catalysed conversion of oxime tosylates to α-aminoketones (3.2). This highly strained structure survived a critical re-examination by Cram twenty years later. The lone-pair which would be sp^2 hybridised in an acyclic imine is given more s character by being on a three-membered ring imine (see § 2.1 and § 2.3) and so the base strength is correspondingly reduced, approaching that of a nitrile. 2-Phenyl-3-methyl-1-azirine is insoluble in 10 per cent hydrochloric acid.

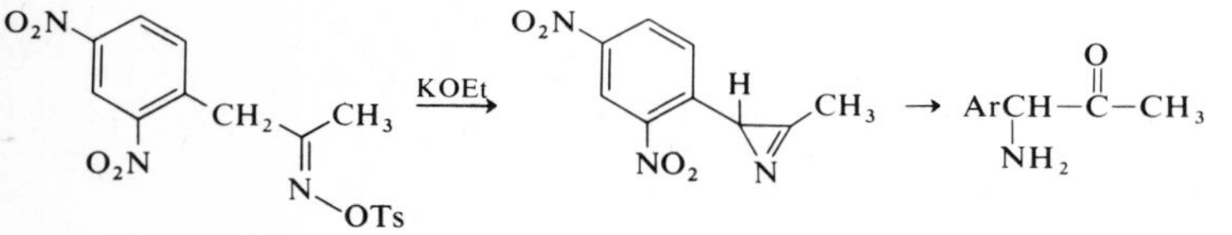

(3.2)

§ *3.1.2   Oxirene*

This should, by analogy to 2-azirine, be a resonance-destabilised 4π electron system. Many reported syntheses of oxirenes have been disproven, although it is thought that oxirenes may be transient intermediates in the oxidation of alkynes with peracid (3.3) and in the decomposition of α-diazoketones (3.4).

Stable exomethylene oxiranes substituted by bulky groups can be synthesised from allenes by peracid oxidation as in (3.5).

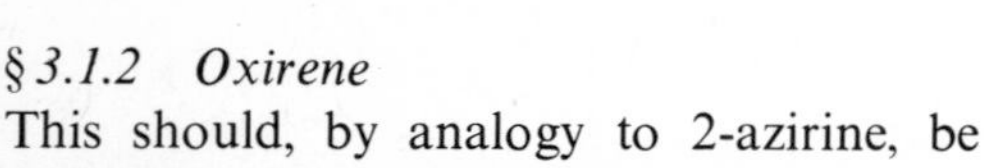

(3.3)

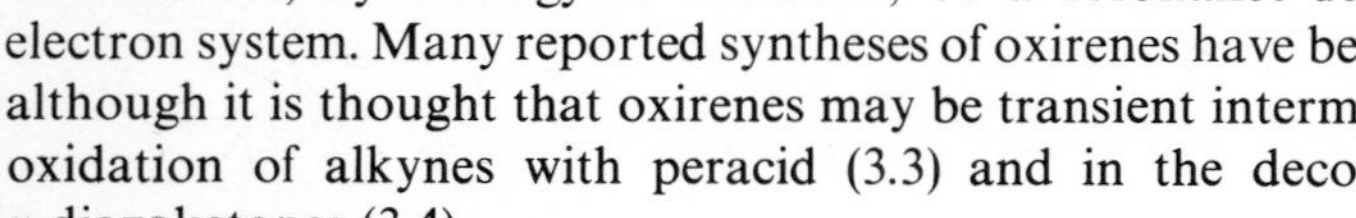

(3.4)

$$\underset{\substack{\text{'Bu} \\ \text{H}}}{\text{H}}\text{C}=\text{C}=\text{C}\underset{\substack{\text{'Bu}}}{\overset{\text{H}}{}} \xrightarrow{\text{RCO}_3\text{H}} \quad \underset{\text{'Bu}}{\overset{\text{H}}{}} \text{O} \quad \xrightarrow{100°\text{C}} \quad \text{'Bu} \quad \text{'Bu} \qquad (3.5)$$

§ 3.1.3　Diazirine

The diazirine structure (**11**) was first suggested for diazomethane, since it was the only classical valence-bond structure which could be written for this molecule. The linear structure (**12**) was eventually preferred on the basis of spectroscopic studies. Strained cyclic diazirines were eventually prepared by oxidation of diaziridines as in (3.6) and these compounds were quite different from the corresponding substituted diazomethanes and could be reduced as in (3.6) and reacted with Grignard reagents as in (3.7) without fission of the ring. Pyrolysis or photolysis yielded carbenes as in (3.8).

$$\underset{(11)}{\text{H}_2\text{C}\overset{\text{N}}{\underset{\text{N}}{|}}} \qquad \underset{(12)}{\text{CH}_2=\overset{\oplus}{\text{N}}=\overset{\ominus}{\text{N}}: \ \leftrightarrow\ \overset{\ominus}{\text{CH}_2}-\overset{\oplus}{\text{N}}\equiv\text{N}: \ \leftrightarrow\ \overset{\oplus}{\text{CH}_2}-\text{N}=\overset{\ominus}{\text{N}}:}$$

$$\underset{\text{R}'}{\overset{\text{R}}{}}\overset{\text{NH}}{\underset{\text{NH}}{|}} \quad \underset{\text{Na}}{\overset{\text{HgO}}{\rightleftharpoons}} \quad \underset{\text{R}'}{\overset{\text{R}}{}}\overset{\text{N}}{\underset{\text{N}}{||}} \qquad (3.6)$$

$$\underset{\text{R}'}{\overset{\text{R}}{}}\overset{\text{N}}{\underset{\text{N}}{||}} \quad \xrightarrow[\text{(2) H}_2\text{O}]{\text{(1) R}''\text{MgX}} \quad \underset{\text{R}'}{\overset{\text{R}}{}}\overset{\text{N}-\text{R}''}{\underset{\text{NH}}{|}} \qquad (3.7)$$

$$\underset{\text{R}'}{\overset{\text{R}}{}}\overset{\text{N}}{\underset{\text{N}}{||}} \quad \xrightarrow{\Delta\,\text{or}\,h\nu} \quad \underset{\text{R}'}{\overset{\text{R}}{}}\text{C}: \quad \xrightarrow{} \quad \underset{\text{R}'}{\overset{\text{R}}{}} \qquad (3.8)$$

§ 3.1.4　"Aromatic" heterocyclic compounds with 2π-electrons

Thiirene-1,1-dioxide (**13**) is analogous to cyclopropenone (**14**) which, in its dipolar resonance form (**14b**), is a planar cyclic conjugated 2π-electron system and so has resonance stabilisation. The similarity of the nmr spectra of the methyl derivatives (**15**) and (**16**) and the thermal stability of diphenylthiirene-1,1-dioxide (**17**) support the idea that this system is stabilised, although solvolysis studies suggest that the degree of stabilisation may not be great.

Where the heteroatom has a vacant p orbital (eg. B, Al, Ga, etc.) or vacant d orbitals (Si, Ge, Sn, etc.) one might expect the heterocyclic compound of general structure (**18**) to be a 2π-electron aromatic compound. In the early 1960s some Russian work suggested that silicon and germanium analogues were obtainable and had considerable stability but mass spectroscopic and X-ray studies have proved that the structure thought to be (19) was in fact the dimeric (20).

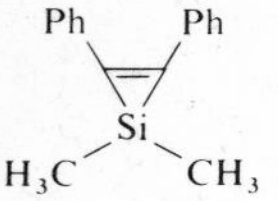

(19) 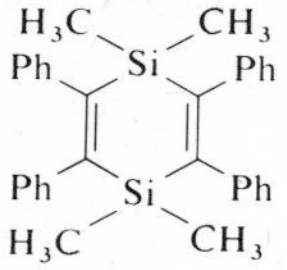   (20)

(3.9)

§ 3.1.5 *Azetes and azetidines*

Fully unsaturated four-membered heterocyclic compounds such as azetes (21) and diazetes (22) are of interest, being iso-π-electronic with cyclobutadiene. They should, therefore, be destabilised by delocalisation. There is no proven report of such compounds being made but the unsaturated lactam (23) prepared by Henery-Logan in 1963 would be expected to give such an antiaromatic structure if it were capable of amide resonance to (23b) and one might expect atypical properties for this amide. Very little chemical and physical data have, however, been reported for this compound.

(21)   (22) 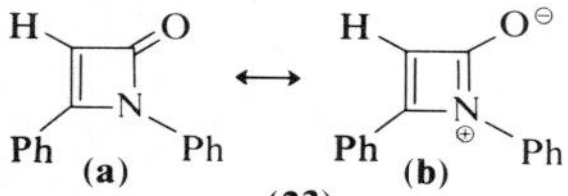   (23)

(3.10)

The monounsaturated compounds such as (24), (25) and (26) can be prepared without much trouble.

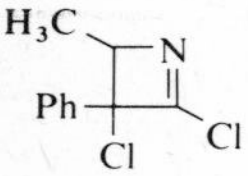

(24) 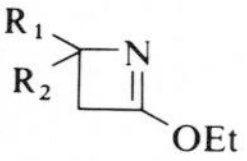   (25)    (26)

§ 3.1.6 *Oxetes*

These are very strained enol ethers and readily rearrange to enones on heating as in (3.11) and (3.12). They may be reduced to oxetanes (3.12) and two methods of synthesis are shown in (3.11) and (3.12).

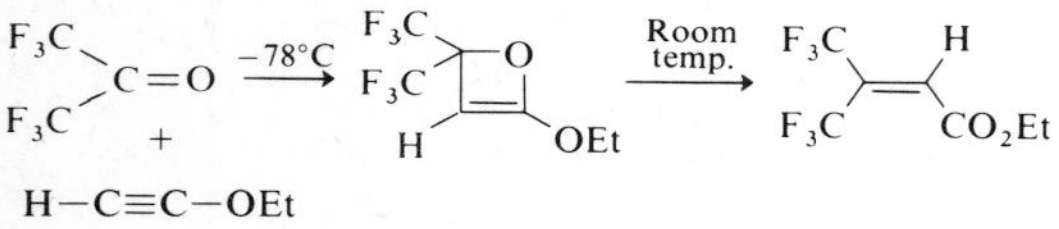

(3.11)

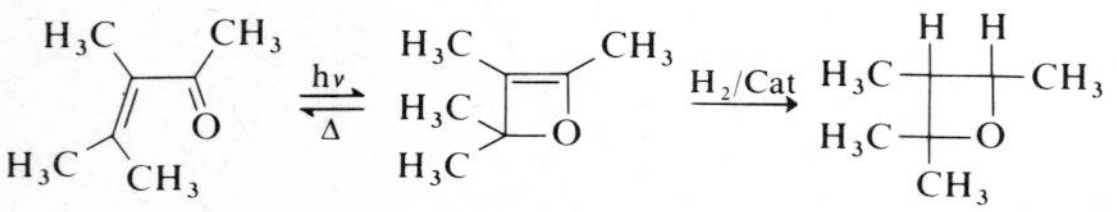

(3.12)

§ 3.1.7 *Thiete* (28)

This was at one time a compound of some theoretical interest since the anion (27) would be a 6π-electron system, and therefore capable of resonance stabilisation. When the parent thiete (**28**) was synthesised, however, it did not appear to be especially acidic. Thietes hydrolyse with ring opening, polymerise, and are oxidised to sulphones.

(**27**)   (**28**)   (**29**)

Dithiete (**29**) might be considered to be a 6π-electron system but when a substituted dithiete was synthesised, it showed mainly aliphatic properties.

§ 3.1.8   Small bicyclic enamines

These, having the nitrogen at a bridgehead, are quite unlike acyclic or even other cyclic enamines since resonance forms such as (**30b**) below are precluded by Bredt's Rule. This makes enamines such as the quinuclidene (**30**) atypical in their reactivity and this compound is not hydrolysed by dilute acid.

(**a**)   (**b**)
        (**30**)

§ 3.2   N-Acylaziridines

Although there is formally no unsaturation in the heterocyclic ring of N-acylaziridines (**31a**), normal amide resonance would imply some contribution from (**31b**). This resonance form would involve an sp^2 hybridised atom in the ring and so one would expect less contribution from (**31b**) to the final hybrid than in acyclic or larger ring amides. N-Acylaziridines have, therefore, more ketone- and amine-like properties than their acyclic or larger ring counterparts and the carbonyl absorption in the infra-red spectrum is more typically that of a ketone than of an amide.

The unique properties of N-acylaziridines have been put to good use in aldehyde synthesis. Moderately hindered amides can be reduced with lithium aluminium hydride to amines as in (3.13) and the preponderance of form (**31a**) makes *N*-acylaziridines more prone to attack by nucleophilic reagents. Further, step (**a**) would involve formation of the sp^2 hybridised aziridine (**33**) and so would be disfavoured. The intermediate (**32**) would therefore survive until quenched with water, when aldehyde would result, as in (3.14) below.

(**a**)   (**b**)
        (**31**)

$$\text{RC} \overset{\displaystyle O}{\underset{\diagdown N \triangleleft}{}} \xrightarrow{\text{LiAlH}_4} \left[\underset{(32)}{\overset{\displaystyle \text{OAlH}_3}{\underset{\diagdown N \triangleleft}{R-\overset{|}{\underset{|}{C}}-H}}} \right]^{\ominus} \xrightarrow{a} \left[\underset{(33)}{\text{RCH} = \overset{\oplus}{N} \triangleleft} \right] \xrightarrow{\text{LiAlH}_4} \text{RCH}_2\text{N} \triangleleft \qquad (3.13)$$

$$\left[\underset{(32)}{\overset{\displaystyle \text{OAlH}_3}{\underset{\diagdown N \triangleleft}{R-\overset{|}{\underset{|}{C}}-H}}} \right]^{\ominus} \xrightarrow{\text{H}_2\text{O}} \left[\overset{\displaystyle O^{\ominus}}{\underset{\diagdown N \triangleleft}{R-\overset{|}{\underset{|}{C}}-H}} \right] \rightarrow \text{RCH} = O \qquad (3.14)$$

§ 3.3   Lactams

§ 3.3.1   α-Lactams

These were postulated as reactive intermediates as early as 1908 but the ring strain in such structures made them elusive until 1961 when Baumgarten detected a compound in reaction (3.15) which had carbonyl absorption at 1847 cm^{-1} in the infra-red. By careful work in inert solvents, he was able to isolate the α-lactam and prove its structure. Many α-lactams have since been made, and some, such as the 1,3-ditertbutyl α-lactam (**34**), are surprisingly stable. This stability is probably due to the Van der Waals repulsive interactions illustrated in (**34**) which oppose ring-strain forces, and to the steric repulsion to nucleophilic reagents of the *tert*-butyl groups.

$$\underset{}{\text{PhCH}_2\overset{\displaystyle O}{\overset{||}{C}}-\text{NH}^t\text{Bu}} \xrightarrow{{}^t\text{BuOCl}/{}^t\text{BuOK}} \left[\text{Ph}\overset{\ominus}{\underset{}{C}}\text{H}-\overset{\displaystyle O}{\overset{}{C}}\underset{\underset{Cl^{\ominus}}{\overset{|}{N}}}{\diagdown}{}^t\text{Bu} \right] \rightarrow \underset{{}^t\text{Bu}}{\overset{\text{Ph}}{\diagdown}\overset{}{\underset{\overset{|}{N}}{\text{CH}}}\overset{\displaystyle O}{\diagdown}} \qquad (3.15)$$

(**34**)

Because of ring strain, α-lactams will have inhibited amide resonance (**35a**) ↔ (**35b**), and the expected lack of contribution from the resonance form (**35b**) is reflected in the high carbonyl absorption in the infra-red spectrum. α-Lactams are keto-aziridines rather than amides and, since both cyclopropanones and aziridines (cf. § 2.6) react with nucleophilic reagents, it is of interest to see how α-lactams react with nucleophiles. Protic nucleophiles tend to yield the product of alkyl-nitrogen fission, *i.e.* the product expected of a "pull" mechanism for ring opening of epoxides (see § 2.6), as in (3.16), while aprotic nucleophiles attack at the carbonyl atom and then undergo ring fission as in (3.17).

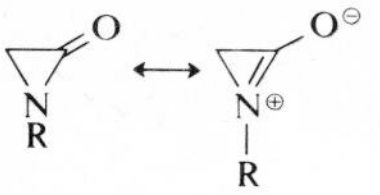

(a)   (b)
(35)

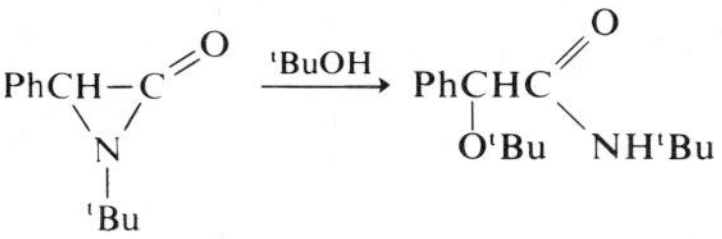

PhCH—C=O →('BuOH)→ PhCHC=O
  | | \
  N   O'Bu   NH'Bu
  |
  'Bu

(3.16)

PhCH—C=O →('BuO⁻)→ PhCH—C(O⁻)(O'Bu) → PhCH—CO'₂Bu
  | | |
  N   N   NH'Bu
  | |
  'Bu   'Bu

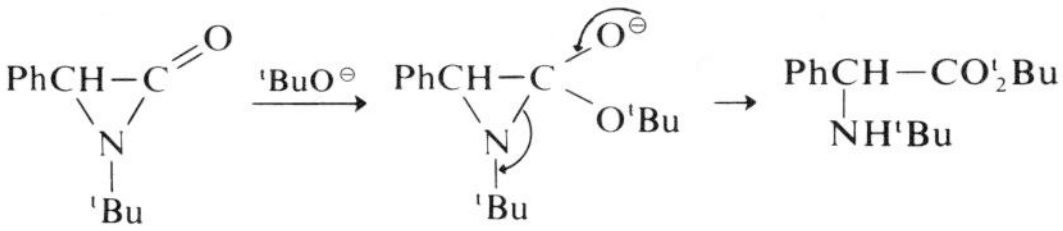

(3.17)

§ 3.3.2   β-Lactams

These have been known since 1907 but it is only since the finding that the important penicillin series of antibiotics (see § 8.2) had the β-lactam-containing structures **(36)** that these systems have been studied extensively. A variety of synthetic methods including cyclisations such as (3.18) and cycloadditions such as (3.19) have been used to make β-lactams.

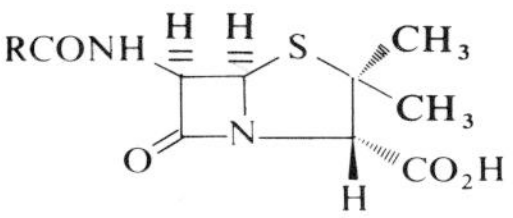

(36)

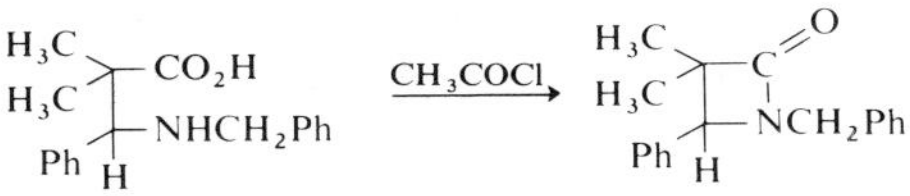

(3.18)

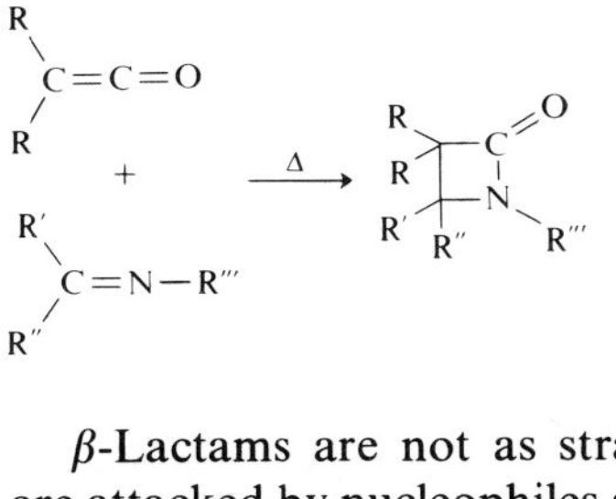

(3.19)

β-Lactams are not as strained as α-lactams and, like most amides, are attacked by nucleophiles at the carbonyl atom. Unlike normal acyclic amides or lactams, however, there is ring strain and the tetrahedral intermediate decomposes with ring fission as in (3.20), (3.21) and (3.22).

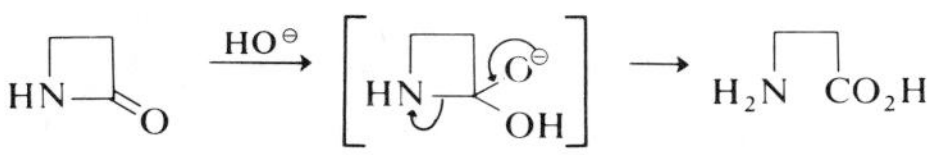

(3.20)

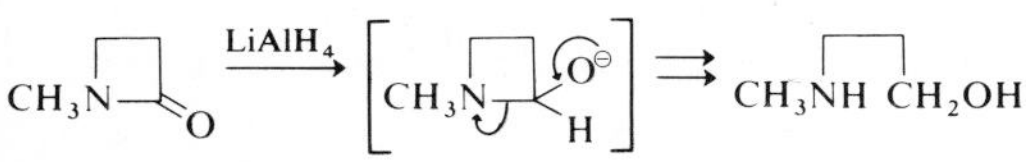

(3.21)

(3.22)

§3.3.3   Larger ring lactams

These are less strained than α- and β-lactams and behave much like acyclic amides. If secondary they can have contributions from the geometrically distinct dipolar forms (37b) and (c) which will be separated by a definite energy barrier. X-ray experiments in the solid state and dipole and infra-red experiments in solution suggest that the *trans* form (37c) is the more stable form for acyclic amides and large-ring lactams. When the amide is constrained in a medium-sized ring it is, however, forced into the unstable *cis* form. This change in geometry is reflected in some of the physical properties of the lactams and in §3.5.2 we shall see how this geometric effect can affect the chemical reactivities of lactones.

(37a)   (37b)   (37c)

§3.3.4   Bridged lactams

Quinuclidone (38), which has a bridgehead nitrogen, does not have normal amide properties since the resonance form (38b) would violate Bredt's rule. This results in the basicity of quinuclidone being more akin to that of an amine and the carbonyl absorption in the infra-red spectrum being 80 cm^{-1} above that for normal amide absorption. Quinuclidone may form oximes and readily undergoes deuterium exchange on the carbon atom α- to the carbonyl group. Such bridgehead amides are very readily hydrolysed.

(a)   (b)

(38)

§3.4   Cyclic ureas

Diaziridones may be synthesised and bulky groups will stabilise the normally strained ring by steric repulsion. 1,2-Di*tert*butyl-diaziridone has a higher than usual carbonyl absorption and lower than usual dipole moment, indicating once again the suppression of the amide resonance form (39b) by ring strain.

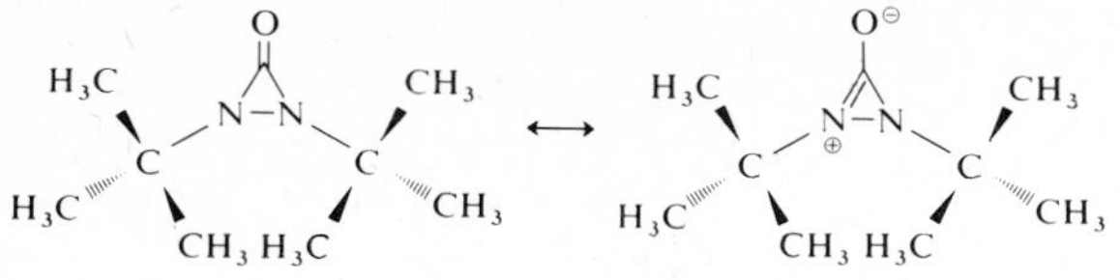

(39)

Nucleophilic attack is invariably at the acyl carbon in the case of these ureas:

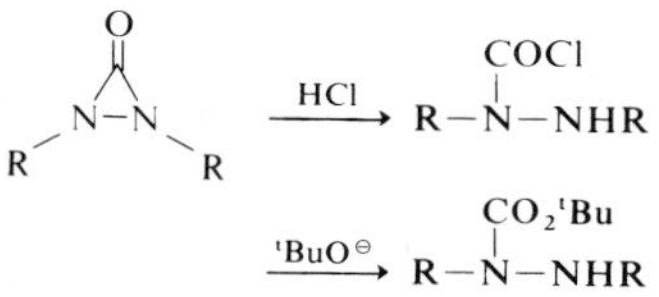

(3.23)

(3.24)

§ 3.5 Lactones

§ 3.5.1 α-Lactones

These were postulated as reactive intermediates in the 1930s but it is only very recently that they have been synthesised and isolated. The first such synthesis by Bartlett in 1970 involved ozonolysis of a ketene (3.25). Bulky *tert*-butyl groups were used to stabilise the strained ring (cf. § 3.3.1) but the α-lactone was not stable above $-20°C$. As might be expected from § 3.3.1 and § 2.6 alkyl-oxygen cleavage seems common here and the carbonyl absorption is at *ca.* $1900\,cm^{-1}$ in the infra-red spectrum.

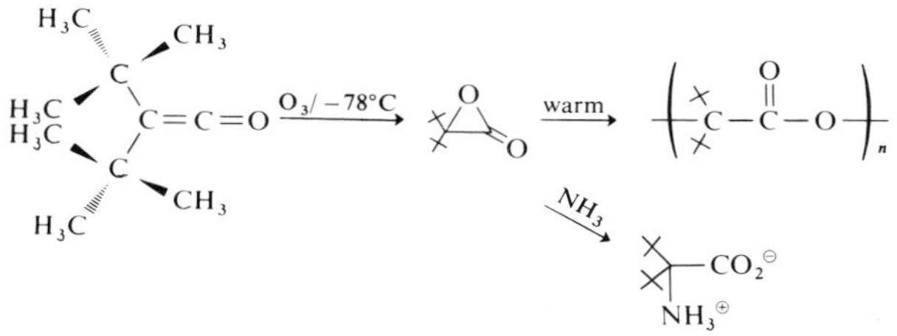

(3.25)

§ 3.5.2 Higher lactones

β-Lactones have been known since 1883 and can be made by cyclisation (3.26) or cycloaddition reactions (3.27).

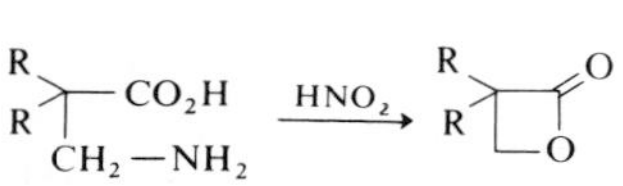

(3.26)

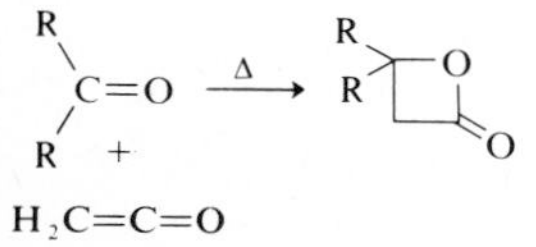

(3.27)

The infra-red absorption of β-lactones is at *ca.* 1840 cm^{-1}. β-Lactones and medium-ring lactones with up to seven atoms in the ring are atypical and, unlike acyclic esters and higher lactones, undergo oxygen–alkyl cleavage in reaction with some anionic nucleophiles such as Cl$^-$, CN$^-$ and RS$^-$ (cf. (3.28)). Larger-ring lactones and esters invariably undergo oxygen-acyl fission. The course of the reaction of β-lactones with ammonia and amines depends on solvent as shown in (3.29) and (3.30).

$$\text{O}\diagdown\text{O} \xrightarrow{\text{NaX}} \quad \text{X} \quad \text{CO}_2{}^{\ominus} \tag{3.28}$$

$$\text{O}\diagdown\text{O} \xrightarrow{\text{NH}_3/\text{Protic solvents}} \quad \text{HO} \quad \text{CONH}_2 \tag{3.29}$$

$$\xrightarrow[\substack{\text{Aprotic} \\ \text{solvents}}]{\text{NH}_3} \quad \text{H}_2\text{N} \quad \text{CO}_2\text{H} \tag{3.30}$$

Huisgen has studied the rates of hydrolysis of the homologous series of lactones **(40)** from n = 4 to n = 14 and correlated these with dipole moment and boiling point as shown in Table 3.1.

$$(\text{CH}_2)_{n-2}\diagup^{\text{C=O}}_{\text{O}} \xrightarrow{\text{H}_2\text{O}} \begin{array}{c}(\text{CH}_2)_{n-2}-\text{CO}_2\text{H} \\ \text{OH}\end{array} \tag{3.31}$$

(40)

Table 3.1

n	Boiling point (°C)	μ(D)	10^4 K
4	51	3·8	—
5	79	4·09	1480
6	97	4·22	55 000
7	104	4·55	2550
8	80	3·70	3530
9	72	2·25	116
10	86	2·01	0·22
11	100	1·88	0·55
12	116	1·86	3·3
13	130	1·86	6·0
14	143	1·86	3·0
16	169	1·86	6·5
Acyclic ester		1·79	8·4

The lactones seem to be in two distinct groups. Those from n = 4 to n = 7 hydrolyse much more quickly than acyclic esters and have high dipole moments. Those from n = 10 to n = 16 hydrolyse much more slowly and have dipole moments of the same order as acyclic esters. Within each series the boiling point increases linearly with increasing molecular weight. Lactones of n = 8 and n = 9 are intermediate between the two series. There is also a tendency, as we have seen, for the series $n = 4$ to $n = 7$ to undergo reaction with nucleophiles with oxygen–alkyl cleavage, quite unlike the acyclic esters or the $n = 10$ to $n = 16$ series.

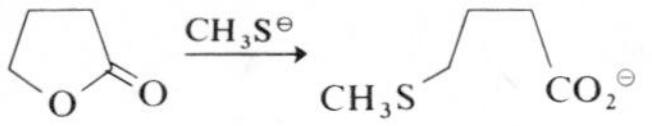

$$\text{(3.32)}$$

The reactivity of the $n = 4$ to $n = 7$ series has been explained as being due to the dipolar resonance form being forced into the *cisoid* configuration (**41b**) by the ring and therefore, like the corresponding amides (cf. § 3.3.3), being less stable than the higher lactones or acyclic esters which can adopt the more stable *transoid* form. Hence the $n = 4$ to $n = 7$ series is more reactive. The molar polarisability figures are in agreement with the postulated geometry of the dipolar form and follow very similar trends to the polarisability of the amides.

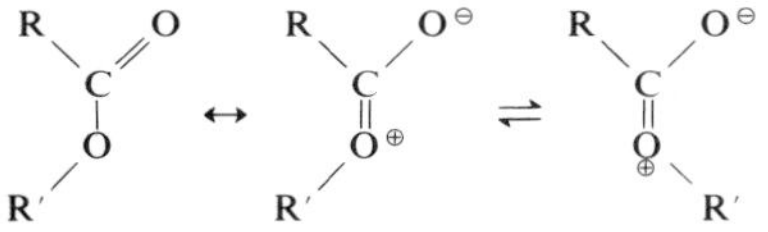

(a)   (b)   (c)

(41)

§ 3.6 Medium ring heterocyclic ketones

In studying the chemistry of certain alkaloids which had a heteroatom (nitrogen) and a carbonyl group arranged across a ring from each other so that they might interact through space, Sir Robert Robinson noted that the carbonyl absorption in the infra-red spectrum showed some amide character. He suggested that such compounds (e.g. cryptopine (**42a**)) might exhibit a "transannular amide resonance" with form (**42b**) having some significance.

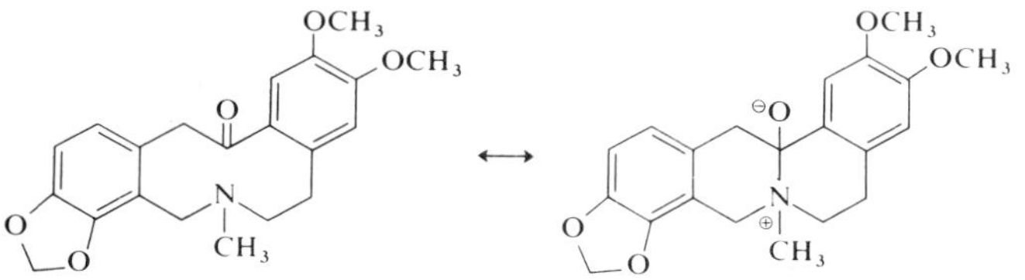

(42a)   (42b)

Subsequent studies of a large series of synthetic heterocyclic ketones of general structure (**43**) have shown that, although there is little interaction of the heteroatom lone-pair for rings smaller than seven and larger than eleven; eight-, nine- and ten-membered rings all exhibit transannular interactions. The eight-membered ring series (**44**) shows this effect very strongly and, as expected, the effect falls off from the less electronegative (and more basic) nitrogen to oxygen, with $N > S > O$. It is most apparent in polar solvents where the dipolar form is stabilised.

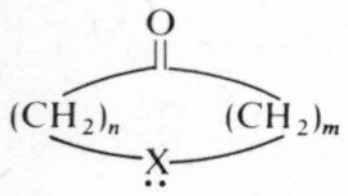

(43) 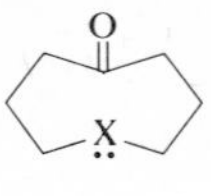   (44) 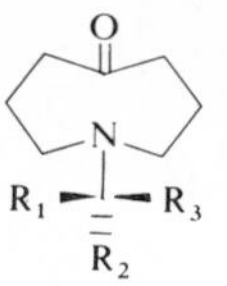   (45)

Evidence for this effect is shown by a second band appearing in the infra-red or charge transfer bands in the ultra-violet. Optically active amines such as (45) can have abnormal optical effects.

Key points

(1) Ring strain effects are dramatically increased by the addition of unsaturated centres to small-ring heterocyclic compounds.
(2) The chemistry of small-ring unsaturated heterocyclic compounds is also affected by such factors as aromaticity, anti-aromaticity and inhibition of resonance.
(3) Bridged heterocyclic compounds with the heteroatom at the bridgehead have properties which are profoundly affected by compliance with Bredt's Rule.
(4) Medium-ring heterocyclic compounds can exhibit interesting transannular effects.

Further reading

General chemistry
Heterocyclic Compounds with Three- and Four-Membered Rings, ed. A. Weissberger, Interscience, New York, 1964.

α-Lactams
I. Lengyel and J. C. Sheehan, *Angewandte Chemie International Edition*, 1968, **7**, 25.

β-Lactams
Cephalosporins and Penicillins, Chemistry and Biology, ed. E. H. Flynn, Academic Press, New York, 1972.

Larger ring lactones and lactams
R. Huisgen and H. Ott, *Tetrahedron*, 1959, **6**, 253 and references cited therein.

Transannular effects in medium rings
N. J. Leonard and C. R. Johnson, *J. Amer. Chem. Soc.*, 1962, **84**, 3701 and references cited therein.

Chapter 4

Aromaticity and tautomerism in heterocyclic chemistry

§ 4.1 Aromaticity – general introduction

Compounds were originally termed aromatic because of their pleasant aroma and a variety of essential oil products were among the first compounds to be called "aromatic". With the development of accurate combustion analysis, aromatic compounds were recognisable by a relatively low H : C ratio, and benzene was regarded as the parent aromatic compound. As structural theory developed, it became apparent that aromatic compounds occupied a unique position, since, although they were obviously unsaturated, they were relatively inert and had none of the classical reactivity of olefinic compounds. Further, although structure (**1**) was the best description of benzene using classical structural theory, it was not possible to isolate two isomeric *ortho*-disubstituted benzenoids such as (**2**) and (**3**).

(1)   (2)   (3)

The inability of classical bonding theory to accommodate the structure of benzene was eventually settled by considering that the structure was represented neither by (**2**) nor by (**3**), but that it was in fact a hybrid of these, the two nearest approximations. In this approach, known as *resonance theory*, the approximations (**2**) and (**3**) are known as *canonical forms* and are theoretical approximations to the structure of benzene, in no way representing the actual structure.

In benzene both canonical approximations should have equally likely contributions to the final hybrid structure, and so any bond should be an average of a double and a single bond. The experimentally derived bond length of 1·39 Å for benzene is intermediate between that for a single bond (1·46 Å) and that for a double bond (1·34 Å). The averaging of the bonds of benzene in this theory is thought to make for a more stable structure and a measure of this stability, the difference between the bond energy of benzene and that of a theoretical cyclohexatriene molecule, is termed *resonance energy*. Resonance energies may be obtained from experimentally derived heats of combustion and hydrogenation.

Naphthalene has three equally possible canonical forms, (4), (5) and (6), and so we should expect bond C_1-C_2 to have two-thirds double bond character and bond C_2-C_3 to have one-third double bond character. This is again reflected in the experimentally derived bond lengths ($C_1-C_2 = 1.37$ Å; $C_2-C_3 = 1.41$ Å).

(4)   (5)   (6)

Molecular Orbital Theory is a mathematical treatment of bonding but pictorially we may consider π-bonding in benzene to result from overlap of *all* of the *p* orbitals from the six sp^2 hybridised carbon atoms shown in (7). That is, the π-electrons will not be localised in any olefinic bond but there will be three bonding orbitals associated with *all* of the carbon atoms present. Calculations imply that delocalisation makes for stability and values may be obtained for *delocalisation energies* from the calculations. The calculations of Hückel have predicted that any planar, fully cyclically conjugated system will be stabilised by delocalisation when there is a total of ($4n + 2$) π-electrons, *n* being zero or any integer, while a similar system which has ($4n$) π-electrons will be destabilised by delocalisation. This is to say that ($4n + 2$) π-systems will be *aromatic* and ($4n$) π-systems will not. The latter have been termed *anti-aromatic*.

Atomic orbitals
(7)

Since the predictions of Hückel, many non-benzenoid ($4n + 2$) and $4n$ π-systems which may be planar and fully cyclically conjugated have been made and the properties of some of the ($4n + 2$) π-series do show some similarities to those of benzene. Some of these systems are summarised in Table 4.1.

Recently, systems such as homoaromatics, bicycloaromatics and spiroaromatics have been made, but since heterocyclic chemistry has impinged little in this area we shall not consider it here.

§4.2   Properties of aromatic compounds

The usual criterion of aromatic character is how closely the properties of the compound concerned resemble those of benzene. That is to say that although a compound may be polyolefin it should have properties

Table 4.1

	Aromatic systems			Anti-aromatic systems		
n	$4n + 2$			$4n$		
0	2 (8)			—		
1	 6 (9)  (10) 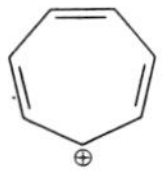 (11)			 4 (12) (13)		
2	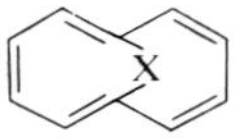 10 (14) 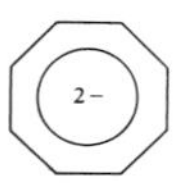 (15) 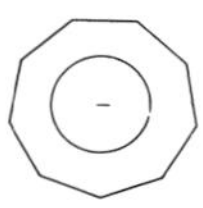 (16)			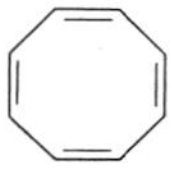 8 (17)  (18)		
3	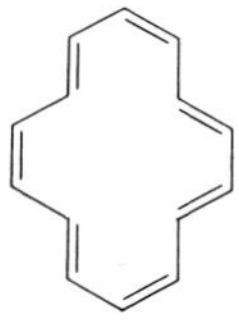 14 (19)					

which are more akin to those of benzene than to those of an olefinic compound. These properties may be summarised as follows.

§ *4.2.1 Physical properties*

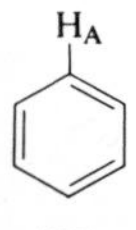

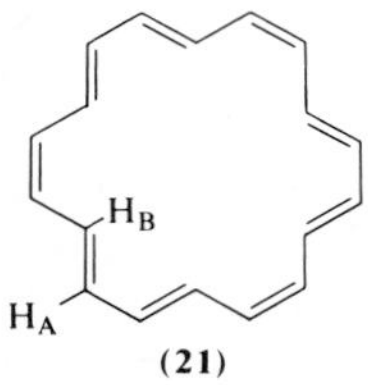

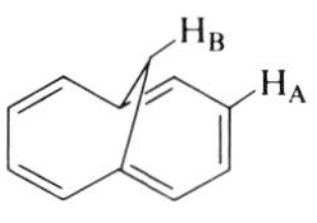

The ultra-violet spectra of aromatic compounds have a distinctive $\pi \rightarrow \pi^*$ absorption and the infra-red spectra have bands which have been ascribed to aromatic vibrations. Dipole moment studies, pK_a, and bond length measurements, and many other physical studies have been used as criteria of aromatic character, as we shall see in the ensuing chapters. One of the most widely used physical methods recently has been nmr spectroscopy. Aromatic compounds with their delocalised electrons set up ring currents, and these ring currents cause nmr deshielding (*i.e.* moving to lower field) of hydrogen atoms which lie in the plane of the ring and to the outside of it. That is to say, H_A in **(20)**, **(21)** and **(22)** is deshielded from the "normal" olefinic chemical shift value of *ca.* 4.5τ to shift values down to 0.72τ. Ring currents also cause nmr shielding (*i.e.* moving to higher field) of hydrogen atoms within the ring, and H_B in (**21**) and (**22**) is shielded to chemical shift values in excess of 10τ.

§ 4.2.2   Chemical properties

The classic property of benzene is its degree of chemical inertness compared with a typical olefin and this lack of reactivity may be taken as a measure of aromatic character. The most characteristic reaction of benzene is its reaction with electrophiles ($X^{\oplus}$ in (4.1) below) to undergo substitution, rather than the addition characteristic of olefins ((4.2) below). Both reactions involve an identical first step but in electrophilic aromatic substitution reactions ((4.1) below), the first formed, "Wheland", intermediate (**23**) regains resonance energy by eliminating a proton. In addition reactions, the positively charged intermediate is quenched by further addition. Electrophilic aromatic substitution reactions are very much influenced by existing substituents on the aromatic ring.

(4.1)

(23)

(4.2)

§ 4.3   Heterocyclic aromatic compounds

Heterocyclic analogues of aromatic carbocyclic compounds may be considered to be of two types. The first type may be obtained by simply substituting a heteroatom for carbon in the aromatic compound, as in the case of pyridine (**24**), pyrimidine (**25**) and quinoline (**26**) below.

(24)   (25)   (26)

These compounds are aromatic by virtue of having $(4n + 2)$ π-electrons delocalised over $(4n + 2)$ atoms and so should be similar to their benzenoid analogues. The benzene system is, however, perturbed by the substitution of the electronegative atom for a carbon atom, and this electronegativity will cause localisation of electron density on the heteroatom(s), resulting in a loss of electron density from the carbon atoms. The chemistry of such compounds is affected by this polarisation of charge and Albert has termed aromatic heterocyclic compounds of this type *π-deficient aromatic heterocyclics* in view of the electron deficiency of the ring carbon atoms. This group of compounds will be dealt with in Chapter 6.

The second type of aromatic heterocyclic compound may be considered to be derived from the carbocyclic analogue by replacement of C=C bonds by heteroatoms as in pyrrole (29) below. The non-bonding lone-pairs on the heteroatoms must now be p hybridised and replace the two π-electrons of the double bond in the $(4n + 2)$ π-system. We now have $(4n + 2)$ π-electrons delocalised over $(4n + 1)$ or less atoms. Since in the canonical forms (27a) to (27e) the heteroatom has "lost" its lone-pair to the carbons of the ring, Albert has referred to such compounds as π-*excessive aromatic heterocyclics*. This group of compounds is dealt with in Chapter 5.

(a)   (b)   (c)   (d)   (e)
(27)

Some heterocyclic aromatic compounds may be considered to be mixed systems, with features of both π-deficient *and* π-excessive compounds in their make-up. Thus imidazole (28) has both a nitrogen replacing a carbon atom, typical of pyridine and other π-deficient systems, and a nitrogen replacing a C=C bond typical of pyrrole (29), a π-excessive system. These compounds will be dealt with in Chapter 5.

(28)   (29)

In the succeeding chapters, we will deal with some of the more common analogues of pyridine and pyrrole, and of polycyclic systems involving six- and five-membered heterocyclic rings. There is, however, considerable interest in less common heterocyclic aromatic (and antiaromatic) compounds and it is appropriate to review some of these systems here.

Looking at Table 4.1, we may readily see that in the row $n = 1$, pyridine (24) may be regarded as a direct analogue of benzene (9) and that pyrrole (29), furan (30) and thiophene (31) could be justifiably considered analogues of the cyclopentadienyl anion (10). Further analogues of the cyclopentadienyl anion (10) would be the thiete anion (32) and dithiete (33), both of which compounds we have considered in Chapter 3 (§ 3.1.7) and neither of which had any especially aromatic properties.

(30)   (31)   (32)   (33)   (34)   (35)

The last of the 6π-carbocyclic aromatic compounds, the tropylium cation (11) has as heterocyclic analogues, the 1,2- or 1,3-dithiolium cations (34) and (35) respectively, with C=C bonds replaced by sulphur atoms which have lone-pairs contributing to the $(4n + 2)$ π-system.

The cations (**34**) and (**35**), which have been prepared, have X-ray-derived bond lengths and nmr spectra which suggest aromatic properties and they are reasonably stable ions.

The antiaromatic compounds with $n = 1$ have been discussed in Chapter 3 and attempts to prepare (**36**), (**37**) and (**38**) have been seen to be abortive. Heterocyclic analogues of the cyclopropenium cation (**8**) such as the sulphone (**39**) and analogues of (**40**) have also been discussed in Chapter 3.

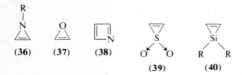

An assortment of 10π-systems such as the azulene (**41**) analogue (**42**) and the cyclononatetraenyl anion analogues (**43**, A = NH, O, S) have been made. The latter compounds have been shown to have nmr spectra compatible with delocalised π-electron systems. The azonines (**44**) and oxonin (**45**) have been made but only the azonines seem to have any claim to aromatic properties. Higher systems than 10π seem, so far, to have no special aromatic properties.

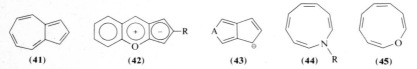

One group of compounds which might be regarded as heterocyclic analogues of the cycloheptatrienyl anion (**18**) are the azepines (**46**), oxepins (**47**) and thiepins (**48**) which, if planar, should be anti-aromatic. Such compounds have been made in recent years and X-ray crystallography has shown that they are not planar and hence are polyenes. There was a feeling that thiepin-1,1-dioxide (**49**), like thiirene-1,1-dioxide (**39**), might be an aromatic system since the empty $3d$ orbitals of sulphur and the polarisation of the S—O bond might make the compound analogous to tropolone (**50**). When the compound was synthesised, however, it proved to have the properties of a simple triene.

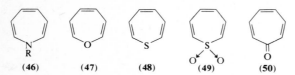

In recent years there has been considerable interest in thiathiophthens which may be considered to have the structure (**51**) below. X-ray studies have shown that the compound is planar, and, if it is symmetrically substituted, both S—S distances are equivalent and larger than an S—S single bond length but shorter than the Van der Waals distance. NMR evidence suggests the presence of an aromatic ring current and so resonance involving forms (51a)–(51b) is suggested. Molecular orbital theory, taking account of empty d orbitals, considers the system to be a delocalised 10π-system with a four-electron three-centre bond over the three sulphur atoms. Selenium may be substituted for sulphur in such compounds but substitution of oxygen for sulphur destroys the unusual properties. An X-ray study on desaurin (52) suggests a planar structure with a short S—O distance implying sulphur–oxygen attraction.

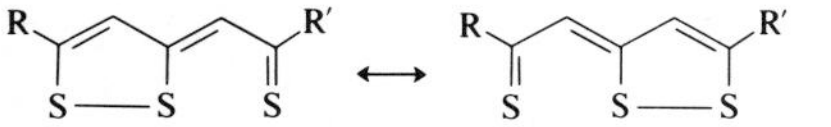

(51a)   (51b) 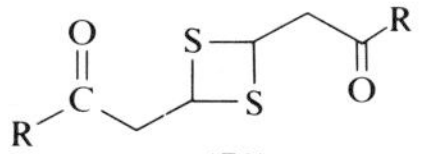   (52)

§ 4.4   Tautomerism in aromatic heterocyclic compounds

Keto-enol tautomerism, the coexistence of, and equilibrium between the keto and enol forms in (4.3), is a well-studied phenomenon. Imine–enamine tautomerism (4.4) and thioketone–thioenol tautomerism (4.5) are equally possible although less thoroughly studied.

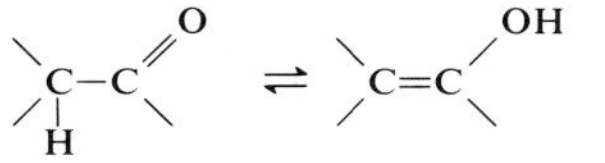

(4.3)

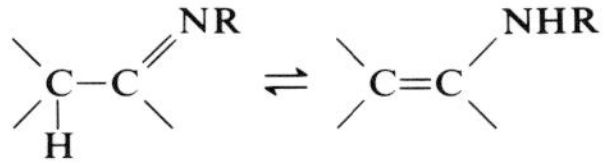

(4.4)

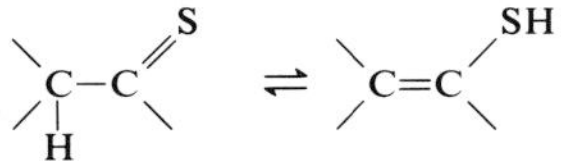

(4.5)

For simple acyclic compounds, the keto form in (4.3) is more stable than the enol form by about $75\,\text{kJ mol}^{-1}$. Thus when we have the possibility, as in (4.6), of the enol form being aromatic and therefore very stable, the equilibrium lies very much to the phenol side. In polyhydroxyphenols, while aromatic stabilisation remains $150\,\text{kJ mol}^{-1}$, each carbonyl group contributes $75\,\text{kJ mol}^{-1}$. Thus phloroglucinol behaves in some reactions as the triketone (54) rather than the phenol (53).

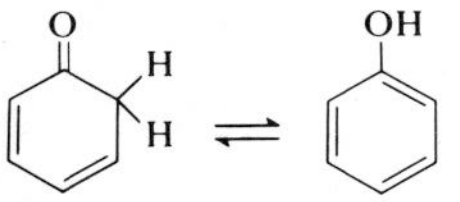

(4.6)

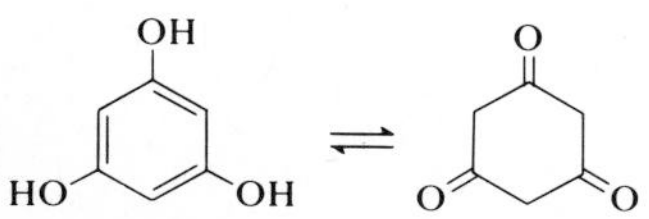

(53)   (54)

2-Hydroxypyridine (**56**) in its "keto" form (**55a**) is not a ketone but an amide and so the amide resonance (**55a**)–(**55b**) will stabilise the "keto" form. Evidence from infra-red, ultra-violet and nmr studies suggests that 2-hydroxypyridine is in fact entirely in the amide form (**55**). The resonance form (**55b**) may be considered as being aromatic and added stability will result from this.

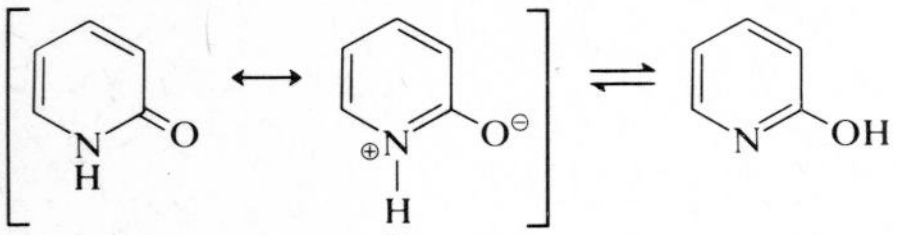

(a) (55) (b) (56) (4.7)

In a similar way, 4-hydroxypyridine (**58**) is, in its amide form (**57a**)–(**57b**), a vinylogous amide. It should, therefore, be stabilised in this form and a variety of physical and chemical experiments have shown that this is indeed the case.

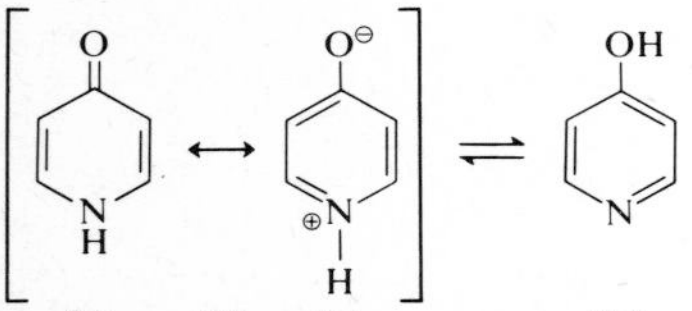

(a) (57) (b) (58) (4.8)

In 3-hydroxypyridine (**62**), the keto forms are in fact simple ketones and not amides. None of the keto forms (**59**), (**60**) or (**61**) has any special stability and so here the phenol form is important. There is, however, evidence of some zwitterionic character (*i.e.* behaviour as in (**63**)) in 3-hydroxypyridines.

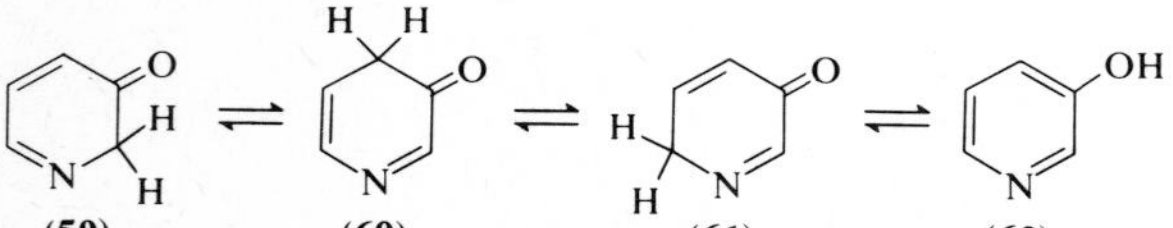

(59) (60) (61) (62) (4.9)

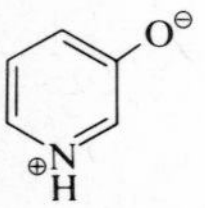

(63)

2,4-Dihydroxypyridine has been shown to exist as 4-hydroxy-2-pyridone (**64**), showing that the fully-conjugated amide form is preferred to the cross-conjugated vinylogous amide form, 2-hydroxy-4-pyridone

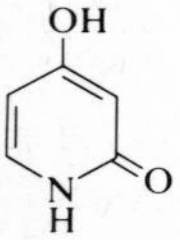

(64)

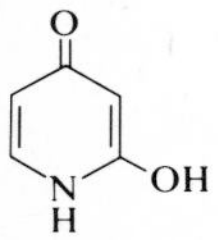

(65)

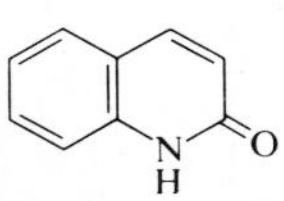

(66)

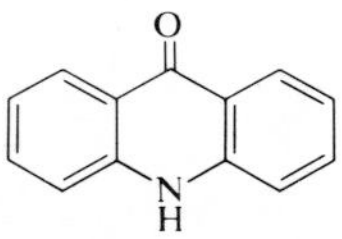

(67)

(65). Fusion of benzene rings to the pyridine system does little to affect tautomerism and so quinolones such as (66) and acridones such as (67) exist in the keto form.

The presence of *additional heteroatoms* makes for a much greater number of tautomers, and so it becomes very much more difficult to assess which tautomer is the most stable in polyheteroaromatic compounds. Each individual case must be investigated on its merits, although in general it is found that the keto form predominates if it is α- or γ- to the heteroatom, just as is the case with substituted pyridines. 4-Hydroxypyrimidines, which are important in natural systems, exist in the amide form (69) and not as (68) or (70). In this respect we have an analogous situation to 2,4-dihydroxypyridine where an amide is preferred to a vinylogous amide. The tautomer (69) also represents the longest chromophore.

(68)   (69)   (70)

It is not difficult to predict now that 2-hydroxypyrimidine exists as the tautomeric form (71); uracil as the tautomeric form (72) (R = H); thymine as (72) (R = CH$_3$); cytosine as (73); barbituric acid as (74); and cyanuric acid as (75). The pyrimidines uracil thymine and cytosine are the bases in nucleic acids and the finding of the correct tautomeric form involved in such compounds played a major part in the unravelling of the double helix structure for DNA (see Chapter 8).

(71)   (72)   (73)   (74)   (75)

Fusion of benzene and heterocyclic five- and six-membered rings to polyheteroaromatic compounds has little effect on the equilibria which have been shown by various studies to be predictable by analogy with the heteromonocyclic compounds above. Examples of such systems are quinazoline-2,4-dione (76); the pteridones (77) and (78); the purine (79); the nucleic acid bases such as (80); and the important redox coenzymes, the flavins (81).

(76)   (77)   (78)

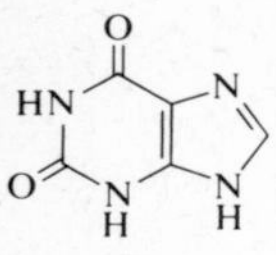

(79) 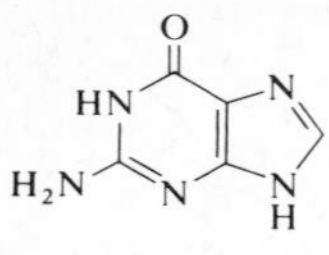   (80) 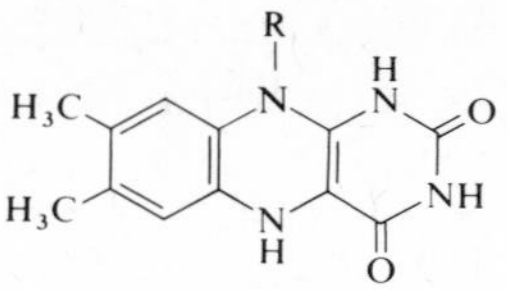   (81)

Six-membered thiophenols and aniline derivatives. On the whole, heterocyclic "thiophenols", like heterocyclic "phenols", exist, when they can,
in the thioamide and thiourea forms, and the series of compounds (**82**)
to (**85**) is completely analogous to the corresponding oxo- series.

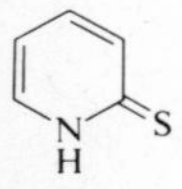

(82) 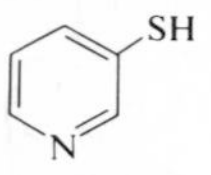   (83) 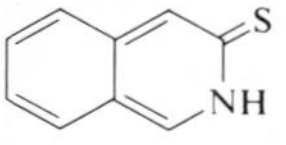   (84) 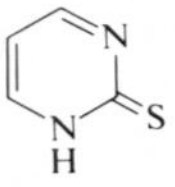   (85)

In the imine–enamine tautomerism (4.4), the imino form is much less
stable than the keto form in (4.3) and so the amino tautomer is often more
favoured than is the case with "phenols". The various compounds (**86**)
to (**89**) are evidence of this fact.

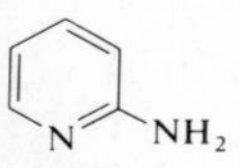

(86) 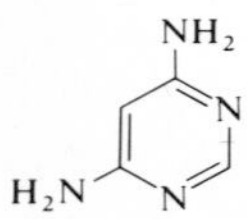   (87) 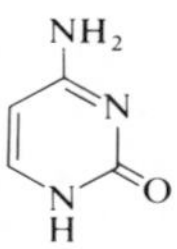   (88) 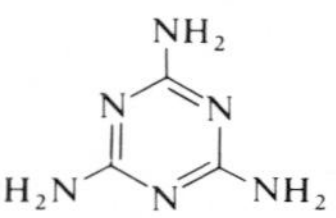   (89)

If the amino group is part of an amide then conjugation may make the
imino form a little easier to obtain, although in (**90**) the enamine form
still predominates. The greater conjugative power of the sulphonamide
group causes the equilibrium to shift to the imine form in (**91**).

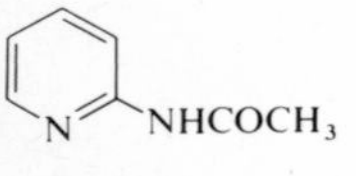

(90) 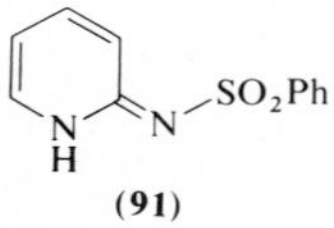   (91)

Five-membered "phenols", "thiophenols" and amino compounds. The
five-membered heterocyclic compounds furan, pyrrole and thiophene
(see chapter 5), if substituted by hydroxyl, thiol or amino groups, can
exhibit tautomerism. Since, as we shall see in Chapter 5, the amount of
aromatic stabilisation in the series falls off in the order thiophene >
pyrrole > furan, it is reasonable to assume that the enolic, and therefore
aromatic, form as in (**92**) or (**95**), is most likely for substituted thiophenes.
Further, since, as we have seen, enamines tend to be preferred forms,
aminothiophenes are the most likely of all of the compounds of this type
to exist in the aromatic forms (**92**) or (**95**).

Three tautomeric forms are possible for 2-substituted compounds as in (4.10) below, and two tautomeric forms are possible for 3-substituted compounds as in (4.11) below.

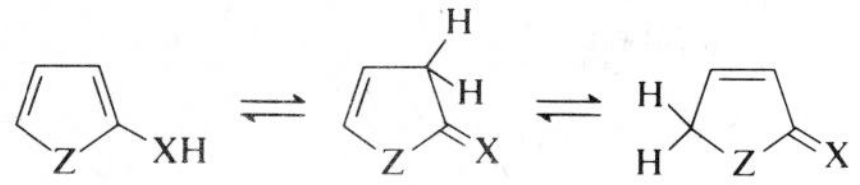

(92) ⇌ (93) ⇌ (94)   (4.10)

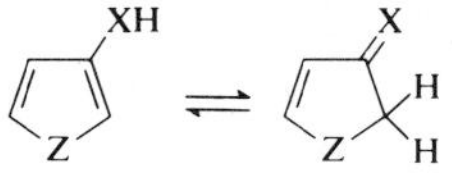

(95) ⇌ (96)   (4.11)

Of the compounds studied, the least likely to exist as the aromatic tautomer is 2-hydroxyfuran, and both of the lactone forms (93) and (94) (X = Z = O) have been observed in the natural products α- and β-angelica lactones. There is evidence that both 2-hydroxythiophene and 2-hydroxypyrrole exist as the "ketonic" tautomers.

The 3-substituted series is less understood than the 2-substituted series although it would appear that 3-hydroxyfuran and 3-hydroxythiophene exist as mixtures of the tautomers (95) and (96). Fusion to a benzene ring with the inherent delocalisation of one of the olefinic bonds by the benzene fragment makes the keto form more likely, and both hydroxyindoles (97) and (98) exist in the keto form. Hydrogen bonding as in (99) can stabilise the "phenolic" tautomer.

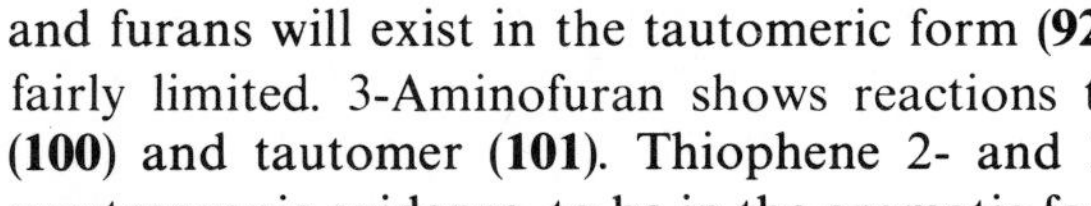

(97)   (98)   (99)

Since the aromatic tautomer is favoured by amino substituted hetero-aromatic compounds, it is expected that 2-amino-pyrroles, thiophenes and furans will exist in the tautomeric form (92) but evidence for this is fairly limited. 3-Aminofuran shows reactions typical of both tautomer (100) and tautomer (101). Thiophene 2- and 3-thiols seem, from nmr spectroscopic evidence, to be in the aromatic forms (92) and (95).

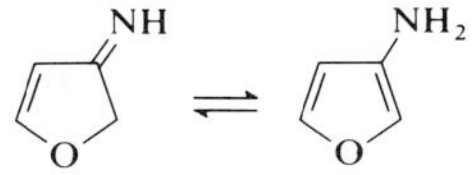

(100) ⇌ (101)

Benzene rings fused β to five-membered heteroaromatic compounds, as in (102) below, tend to favour the keto tautomer as does the presence of a second heteroatom as in saccharin (103), and the compound (104).

Tautomerism in mixed-ring systems such as (105) and (106) can be predicted from a knowledge of the simpler systems.

(102)   (103)   (104)   (105)   (106)

Other tautomeric systems. Systems such as (107) and (108) below with bivalent heteroatoms exist in tautomeric forms which may be predicted on the basis of arguments used elsewhere in this chapter. The problems are simplified somewhat by the fact that the heteroatom is bivalent. Tautomerism such as (109) $\rightleftharpoons$ (110) which can occur in unsubstituted heterocyclic systems will be dealt with in other chapters.

(107)   (108)   (109)   (110)

Key points

(1) Many compounds which have $(4n + 2)$ π-electrons in planar cyclic conjugation may be considered to be aromatic.

(2) Aromatic compounds are compounds with properties resembling those of benzene. They are more stable than non-aromatic polyolefins and have distinctive nmr spectra. Such compounds tend to react with electrophilic reagents in substitution reactions.

(3) Heterocyclic aromatic compounds may be divided into two classes:

(a) π-deficient heteroaromatic compounds,

(b) π-excessive heteroaromatic compounds.

(4) Tautomeric equilibria can be greatly affected by the presence of heteroatoms. π-Deficient heterocyclic aromatic compounds with a hydroxyl group α or γ to the heteroatom do not exist as the phenolic tautomers but as the keto tautomers. Substituted π-excessive heterocyclic aromatic compounds exist in tautomeric forms which are determined by the nature of both the heteroatom and the substituent.

Further reading

Aromaticity

Non-Benzenoid Aromatic Compounds, ed. D. Ginsburg, Interscience, New York, 1959.

Tautomerism in Heterocyclic Compounds

A. R. Katritzky and J. M. Lagowski, *Advances in Heterocyclic Chemistry*, ed. A. R. Katritzky, Academic Press, 1963, vol. 1, 311, 339; vol. 2, 1, 27.

Chapter 5

Five-membered aromatic heterocyclic compounds

§ 5.1 Aromaticity and physical properties

We have seen, in the last chapter, that compounds of the general formula
(1) may be considered to be aromatic and "π-excessive". This is because
the five sp^2-hybridised atoms may sustain a 6-π-electron system as shown
in (5) below. Each carbon atom contributes one p-hybridised electron to
the system and the lone-pair provides the remaining two electrons.
Thus the e are six electrons in five orbitals.

The parent heterocyclic compounds in this series are furan (2), pyrrole
(3) and thiophene (4). Furan and pyrrole involve elements in the first row
of the periodic table and so the available orbitals are much as shown in (5).
Thiophene, however, with its second-row heteroatom sulphur, may
expand its valence shell by use of the empty d orbitals in hybridisation
and so, instead of six electrons in five orbitals, thiophene by using pd^2
hybrid orbitals might have six electrons in six orbitals. This could account
for the extreme stability of thiophene and its great resemblance in
properties to benzene

(1)    (2)    (3)    (4) 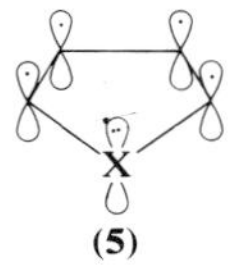   (5)

Valence bond treatment of compounds of general formula (1) considers
them to be hybrids of canonical forms (1a) to (1e), the lone-pair being
"donated" to the ring carbons. When sulphur is the heteroatom, then
forms (6a) to (6e), which imply use of d orbitals, must also be considered.
The controversial tetravalent sulphur here has received some support
recently with the synthesis of the relatively stable compound (7) as in
(5.1) below. This compound appeared to be non-polar and formed an
adduct with N-phenylmaleimide as shown.

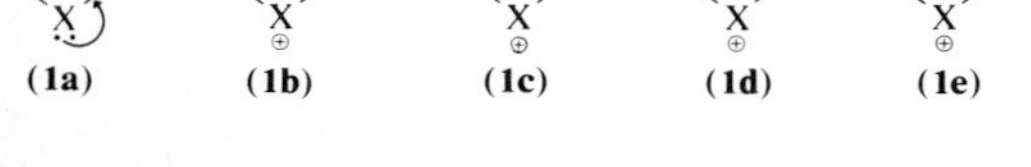

(1a)   (1b)   (1c)   (1d)   (1e)

(6a)   (6b)   (6c)   (6d)   (6e)

(7)

(5.1)

The aromatic nature of compounds in this group is well exemplified by their physical properties. *Bond lengths*, *dipole moments* and *resonance energies* have been measured for the compounds and their tetrahydro derivatives, and these are listed in Table 5.1. The C—X bond length is, in general, shorter in the aromatic compound than in its tetrahydro counterpart, indicating partial double bond character for this bond as is implied by the canonical forms (**1b**) to (**1e**) and (**6a**) to (**6e**). The closeness to unity of the ratio of the C_2—C_3 and C_3—C_4 bond lengths has been taken by some workers as a measure of aromaticity and, using this criterion, the aromaticities of the parent heterocyclic compounds fall in the order thiophene > pyrrole > furan, an order which parallels the calculated resonance energies (which vary in the table according to their derivation).

Table 5.1

| | Bond lengths | | | Dipole moment (D) | Resonance energy (kJ mol^{-1}) | | |
	X—C	C_2—C_3	C_3—C_4		A	B	C
Furan	1·36	1·36	1·43	0·71	96	91	67
Tetrahydrofuran	1·43	1·54	1·54	1·68			
Pyrrole	1·38	1·37	1·43	1·84	130	100	88
Pyrrolidine	1·47	1·54	1·54	1·57			
Thiophene	1·71	1·37	1·42	0·52	130	116	122
Tetrahydrothiophene	1·82	1·54	1·54	1·87			

The dipole moment data listed in Table 5.1 have been assumed by some workers to be evidence for the charged canonical forms (**1b**) to (**1e**) and indeed a systematic study of the dipole moments of substituted pyrroles has suggested that the direction of the dipole in pyrrole is the reverse of that in pyrrolidine. Recent nmr work, however, indicates that the heteroatom is still at the negative end of the dipole in furan and in thiophene but, since the dipole moments of these compounds are smaller than those of the respective tetrahydro derivatives, forms **(1b)** to **(1c)** must counteract the inductive effect of the heteroatom to some extent.

Both molecular–orbital-derived electron densities and the valence bond forms **(1a)–(1e)** support the π-excessive character of these heterocyclic compounds. The aromatic character must be dependent on the availability of the lone-pair for resonance and this in turn will depend on the electronegativity of the heteroatom. More electronegative atoms, such as oxygen, will have a greater "hold" on the lone-pair which will, therefore, be more localised. Furan should therefore be less aromatic than pyrrole or thiophene and the order of aromaticity, thiophene > pyrrole > furan, which we have seen from bond length, dipole and resonance energy studies follows the order of electronegativities (S < N < O).

As we have seen in Chapter 4 (§ 4.2.1), shielding and deshielding characteristics in the nmr spectra of compounds have been used to assess their aromatic character, and attempts have been made to use such data quantitatively. From comparison of such effects, ring currents have always been found to be in the order benzene > thiophene > pyrrole > furan, although the values vary with the method used.

We have so far confined our discussion to three of the monoheteroatomic compounds of general formula **(1)**. In group six of the periodic table the series of compounds **(1, X = O, S, Se, and Te)** has been made, and nmr work has suggested that all of these compounds are aromatic.

In group five of the periodic table, the compounds **(1, X = NR, PR, AsR, SbR)** have been reported. Of this group, only pyrroles and phospholes have been studied in any detail. NMR studies suggest that, in phospholes, the third valency of phosphorus is not coplanar with the ring so that phospholes may not be aromatic. This is in keeping with the pyramidal stability of the lone-pair on phosphorus which we have mentioned in Chapter 2. The fact that phospholes have been shown by variable temperature nmr studies to require less energy for interconversion of the isomer **(8)** to the isomer **(10)** than would a saturated analogue, may indicate that the sp^2 hybrid transition state **(9)** is aromatic.

$$R_1 \quad P \quad R_2 \quad \rightleftharpoons \quad R_1 \quad P \quad R_2 \quad \rightleftharpoons \quad R_1 \quad P \quad R_2$$

| **(8)** | **(9)** | **(10)** |

The group four elements, silicon, germanium and tin, have been reported as participating in five-membered heterocyclic compounds and it has been noted that silole forms a 6π-stabilised anion readily. Pentaphenylborole, which should be destabilised by overlap of the diene system and the empty orbital on boron, reacts, as might be expected, like a diene. Various transition metal analogues of furan have been made recently **(1, X = Cr, Fe, Rh, Ir, Au, Hg, etc.)** and X-ray studies on some of these have not only confirmed structures, but argued for aromatic stabilisation. Apart from the group six heterocyclic compounds and pyrrole, however, the chemistry of this class of compounds has not been thoroughly investigated, and although many claims have been made for the synthesis and aromaticities of the various compounds, we shall avoid this controversial area here.

Since nitrogen is trivalent it may be substituted for a $-CH=$ group in five-membered heterocyclic compounds and so analogues of furan, pyrrole, and thiophene are possible with one, two, three or four nitrogen atoms in addition to the heteroatom already present. The analogues with one additional nitrogen have been termed azoles, and there are six possible azoles (11) to (16) as follows.

(11)	(12)	(13)	(14)	(15)	(16)
Pyrazole	Isoxazole	Isothiazole	Imidazole	Oxazole	Thiazole

The two pyrrole analogues, pyrazole (11) and imidazole (14), when unsubstituted at nitrogen, are capable of the tautomerism shown in (5.2) and (5.3), although this is noticeable only when there is a substituent on carbon to destroy the symmetry of the molecule. Numbering of such compounds is complicated and the atoms are often identified by two numbers to avoid confusion. Thus the substituent in (14a) may be denoted 5(4). This tautomerism is obviously not possible with N-substituted diazoles or the oxygen or sulphur analogues.

$$(5.2)$$

(11a)   (11b)

$$(5.3)$$

(14a)   (14b)

Although it may be convenient to regard the azoles as analogues of furan, pyrrole and thiophene, and these compounds do have many of the properties expected of π-excessive heterocyclic compounds, the doubly bound nitrogen could be considered to be analogous to the nitrogen of pyridine. Some of the properties of azoles are dictated by the presence of this group.

Canonical forms (17a)–(17e) and (18a)–(18e), analogous to (1a)–(1e), may be written for the azoles, and thiazoles and isothiazoles can have further canonical forms analogous to (6a) to (6e).

(17a)	(17b)	(17c)	(17d)	(17e)

$$(18a) \longleftrightarrow (18b) \longleftrightarrow (18c) \longleftrightarrow (18d) \longleftrightarrow (18e)$$

(18a)　　　(18b)　　　(18c)　　　(18d)　　　(18e)

Like canonical forms (**1a**) to (**1e**), the canonical forms for the azoles are not all equivalent in their contribution to the final hybrid. The anions in (**17c**) and (**17d**) are stabilised respectively by the α-C=N and C=X$^{\oplus}$ groups, and so these forms are more important than (**17e**) which has no such stabilisation. As we shall see (§ 5.4), the fact that the preferred site of electrophilic aromatic substitution is C_5 for 1,3-azoles can be explained by such considerations. The stability of anion (**18e**) with two electron-withdrawing α-C=N groups is also reflected in the fact that C_4 is the preferred site of electrophilic substitution in 1,2-azoles, and that substitution often fails if this site is blocked.

Various studies have been made of the physical properties of poly-heteroatomic π-excessive heterocyclic compounds. NMR spectroscopy and other studies suggest that pyrazole and imidazole are aromatic and resonance energies of 123 and 59 kJ mol^{-1} respectively have been obtained for these compounds. Polarisation of the imidazole ring is shown by a dipole moment of about 3·8D. This dipole moment is strongly dependent on the concentration of the solution and this fact, together with the high boiling point of imidazole, has been ascribed to molecular association by facile intermolecular hydrogen bonding as in (**19**).

(19)

§ 5.2　Reactions involving the heteroatom

§ 5.2.1　Basicity

Since pyrrole can form a salt only at the expense of its aromatic character, the lone-pair being essential to the aromaticity of the system, it is a very weak amine. One group of workers has found the pK_a in (5.4) to be -0.27 and it is thus a weaker base than pyridine ($pK_a = 5.2$) with its sp^2 hybridised lone-pair, or aniline ($pK_a = 4.6$) where, although the lone-pair is involved in resonance, it is not essential to maintaining a 6π-electron system. N-Alkyl pyrroles with the electron-donating properties of the alkyl substituent are naturally stronger bases than pyrrole, and electron-withdrawing substituents on the ring will cause the substituted compound to be a weaker base than the parent compound. NMR studies have shown that in fact protonation of pyrrole with acids occurs at a ring carbon atom as in (5.5) below and that this intermediate then trimerises.

$$\underset{H \quad \cdot \quad H}{N} \overset{K}{\rightleftharpoons} \underset{H}{N} + H^+$$

(5.4)

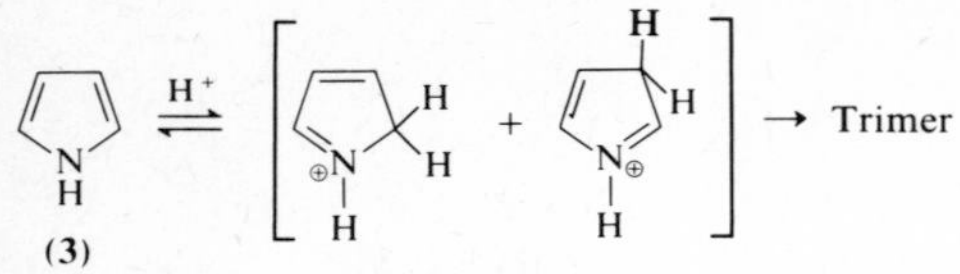

(5.5)

(3)

Furan is readily hydrolysed in acid and this reaction is thought to involve a protonated intermediate analogous to that in (5.5) above. Very mild conditions must therefore be employed if hydrolysis of furan to the dialdehyde (21) is to be achieved rather than polymerisation of the protonated species (20). Thiophene is very stable to acids although very strong acids may bring about oligomerisation.

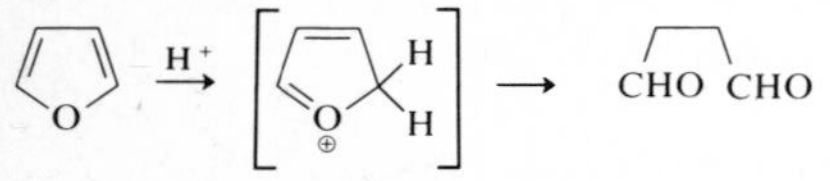

(20)   (21)

The azoles (11) to (16) with their additional pyridine-type nitrogen which is not involved in maintaining aromaticity are understandably more basic than their monoheteroatomic analogues. The basicity of azoles ranges from the strongly basic imidazole ($pK_a = 7$), through thiazole ($pK_a = 2.5$) and pyrazole ($pK_a = 2.5$), to the weakly basic isoxazole ($pK_a = 1.3$). The conjugate acids of the various azoles (e.g. (22) and (23)) are important intermediates in substitution reactions.

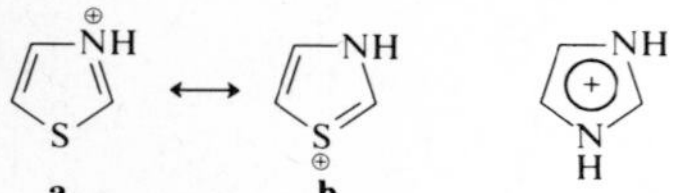

a   b
(22)   (23)

§ 5.2.2 Other reactions involving the lone-pair

N-Substituted pyrroles are not quaternised by excessive alkylation on nitrogen but N-substituted imidazoles, pyrazoles, thiazoles, isothiazoles, oxazoles and isoxazoles may form quaternary salts without destruction of the aromatic sextet, and salts such as (24) to (29) can be formed.

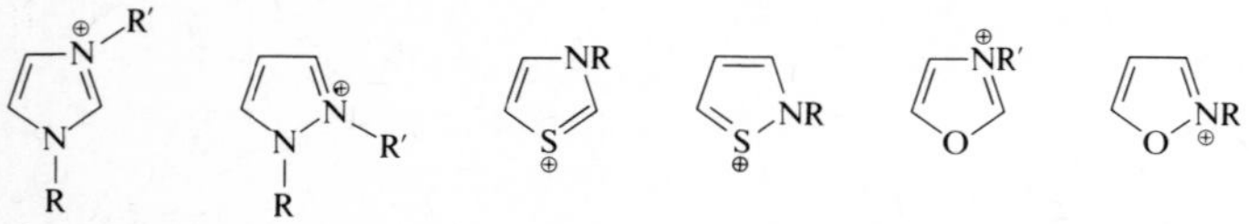

(24)   (25)   (26)   (27)   (28)   (29)

Thiophene may form a stable S-thiophenium salt on treatment with Meerwein's reagent (trimethyloxonium fluoroborate) as in reaction (5.6) and the unstable dioxide (30) has been synthesised and shows diene-like properties (cf. § 5.3).

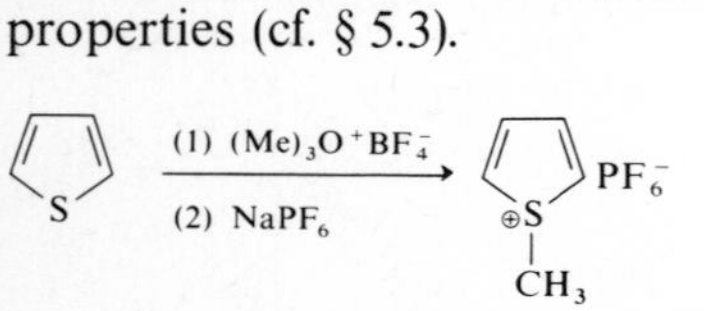

(5.6)

(30)

Hydrogen bonding is especially strong in imidazoles (cf. § 5.1) and the imidazole residue in the amino acid histidine is very important in hydrolytic enzymes, as we shall see in Chapter 8.

§ 5.2.3   Acidic properties

Acidic properties are only relevant to pyrrole and its aza-analogues. Pyrrole with $pK_a = 16\cdot5$ in reaction (5.7) is an acid of comparable strength to methanol and its salts will thus be hydrolysed in water. Its acidity, like that of phenols, may be increased by appropriately placed electron-withdrawing groups and 3-nitropyrrole has an anion stabilised by resonance as in (31) below. A second heteroatom can have the same effect as an electron-withdrawing group as shown in (32), and imidazole and pyrazole are more acidic than pyrrole.

(5.7)

(31)   (32)

Pyrrole forms salts with Grignard reagents and anhydrous bases, the conjugate base being resonance-stabilised as in (33a) to (33e). In this respect it has some resemblance to a phenol where the phenolate anion is stabilised by resonance. Pyrrole, like phenols, can behave as an ambident anion and may be alkylated at either carbon or nitrogen.

(33a)   (33b)   (33c)   (33d)   (33e)

Conditions such as high solvent polarity, reaction homogeneity and low salt concentration tend to favour N-alkylation. The metal ion present also affects the site of alkylation and, for N-alkylation, K > Na > Li. Thus conditions which favour dissociation of the salt favour N-alkylation, whilst C-alkylation is favoured by more associated salts as is the case with phenol alkylation. Acyl halides, ethyl chloroformate, propargyl bromide and many other reagents cause N-alkylation and N-acylation of pyrrole, imidazole, and pyrazole. It is of interest that alkylation of the anion of 2,3,4,5-tetramethyl pyrrole (34) yields the pyrrolenine (35) in which aromaticity has been lost rather than the aromatic 1,2,3,4,5 pentamethyl pyrrole.

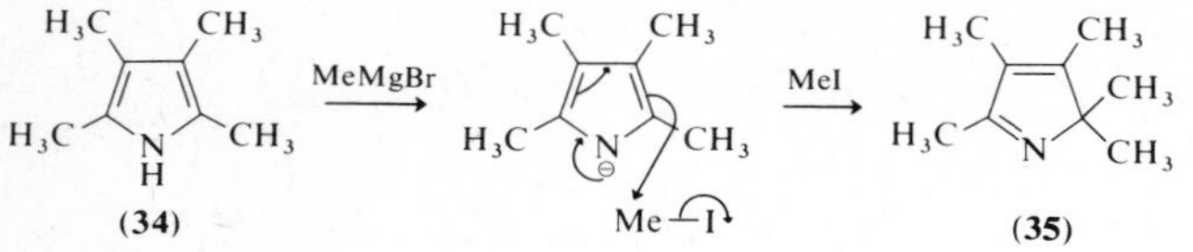

(5.8)

§ 5.3   Reactions typical of dienes

The "aromatic" properties of the five-membered π-excessive heterocyclic compounds cause them to undergo reactions typical of aromatic compounds, such as electrophilic substitution (see § 5.4). The likelihood of such compounds acting as dienes is inversely proportional to the degree of aromatic character they possess and so furan would be expected to be the most "diene-like" compound of the series. Furan reacts with a variety of dienophiles in Diels–Alder reactions such as reaction (5.9). In photochemical reactions furan undergoes [2 + 2] cycloadditions as in reaction (5.10) rather than thermal [4 + 2] cycloadditions such as (5.9). Furan may be oxidised by bromine in methanol to yield the 2,5-diadduct as in reaction (5.11) and there is evidence that some of the electrophilic substitution reactions of furan may proceed via intermediate 2,5-diadducts (see § 5.4).

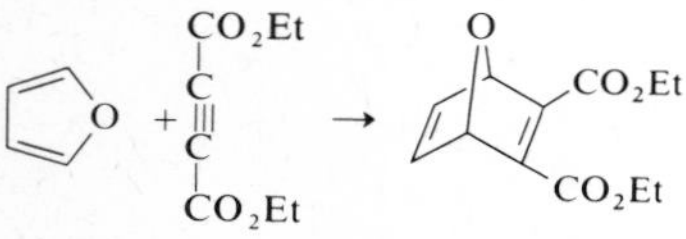

(5.9)

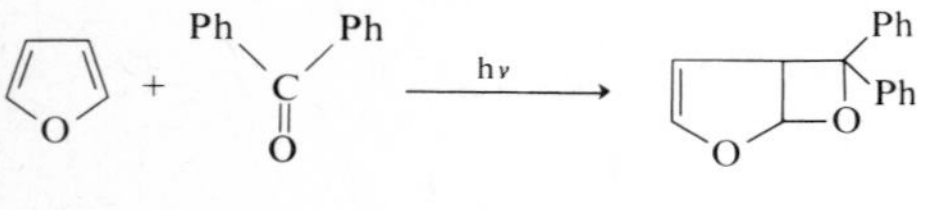

(5.10)

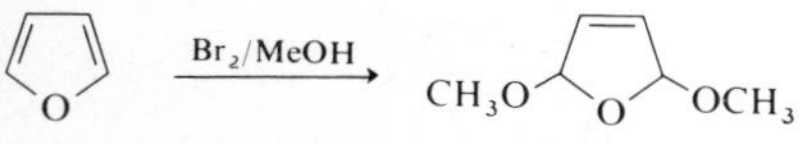

(5.11)

Pyrrole has more aromatic character than furan and so it is less prone to reactions typical of a diene. Since it is basic, however, protonation of the lone-pair can destroy the aromatic character of the compound and the compound will polymerise. Lewis acid catalysts have been found to promote Diels–Alder reactions with pyrroles although very strong dienophiles such as benzyne may react under non-acidic conditions as in reaction (5.12). Substitution on nitrogen with electron-withdrawing groups will reduce the availability of the lone-pair to the ring and so N-carbethoxypyrrole will react with dienophiles as in reaction (5.13). Some dienophiles will actually react as electrophiles in electrophilic aromatic substitution of pyrrole (see § 5.4).

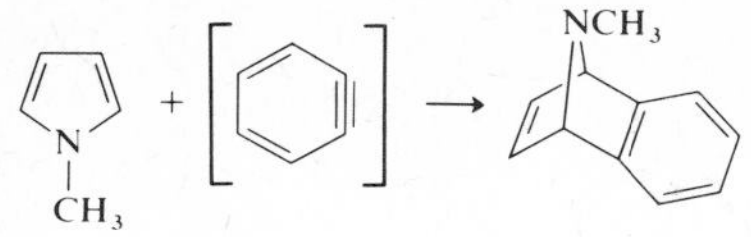

(5.12)

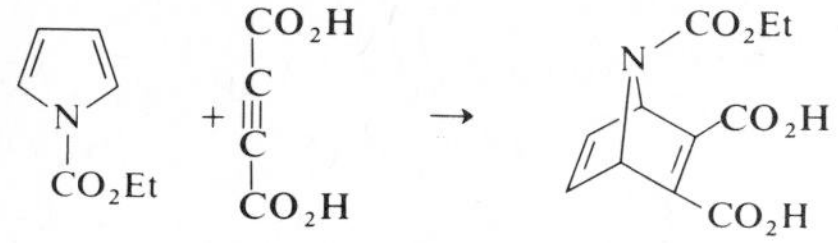

(5.13)

Thiophene, the most aromatic compound of the series, will only occasionally react as a diene. Oxidation to sulphoxides and sulphones in which the "aromatic lone-pairs" are involved in bonding, however, will allow typical diene reactions such as the Diels–Alder reaction (5.14).

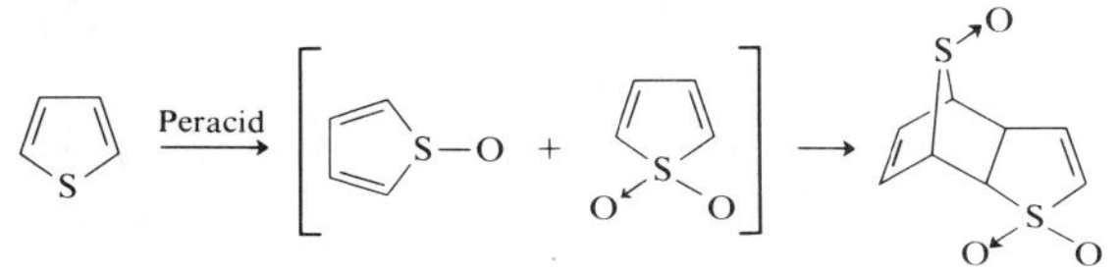

(5.14)

The introduction of a second heteroatom does not affect the diene/ aromatic nature of the series and, not unexpectedly, oxazole behaves much as does furan and reactions such as (5.15) will occur with dienophiles.

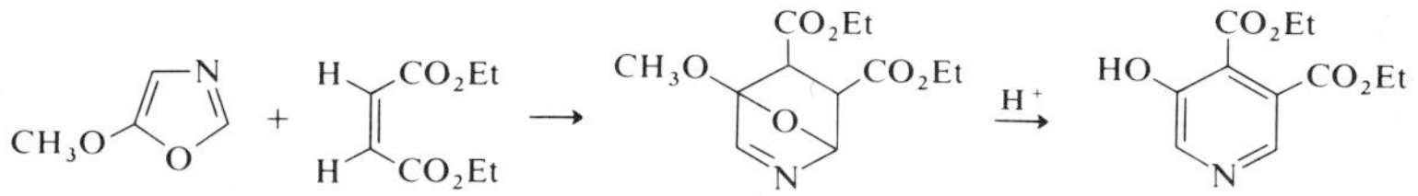

(5.15)

§ 5.4   Electrophilic substitution reactions

The reactivity of benzene is characterised by the ease with which it undergoes substitution with electrophilic reagents. The effect of substituents in directing the electrophile to specific positions on the ring and in altering the rate of reactions of benzenoid compounds has been exhaustively studied. The general mechanism of this reaction has been found to be as in (5.16) below and work on the π-excessive heterocyclic compounds has shown that electrophilic substitution occurs in similar fashion to the reactions of benzenoid compounds. The mechanism outlined in (5.17) and (5.18) has been suggested for these reactions and this involves intermediates similar to the Wheland intermediate (**36**) in (5.16). Since the π-excessive heterocyclic compounds have higher electron density on carbon, it is not surprising that their reactivity towards nucleophilic reagents is higher than that of benzene. They are more akin to benzenoid aromatics substituted by electron-releasing groups.

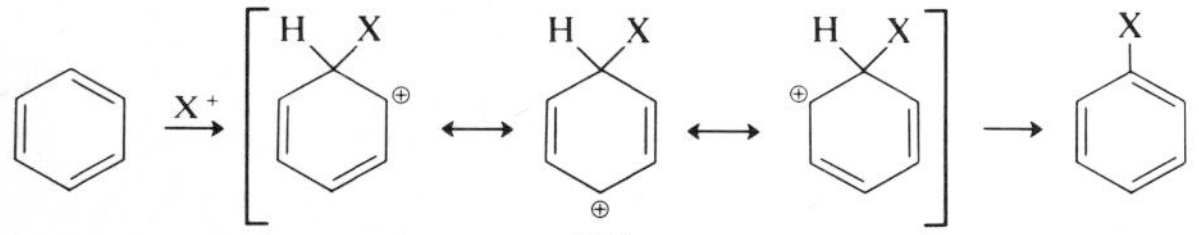

(5.16)

(**36**)

$$(5.17)$$

$$(5.18)$$

Since the Wheland intermediate for 2-substitution in (5.17) has three resonance forms and the intermediate for 3-substitution in (5.18) has two resonance forms, one might expect the former to be more stabilised. It is not surprising, therefore, that substitution at the α (2,5) positions is preferred to substitution at the β (3,4) positions.

The similarity of the reactions, such as electrophilic aromatic substitution, of π-excessive heterocyclic compounds to the reactions of benzene is evidenced by the applicability of Hammett relationships to such reactions (see § 5.6). Rate studies indicate that degree of reactivity is determined more by electron availability than degree of aromaticity and the order of reactivity towards electrophiles is N-methylpyrrole > pyrrole > furan > selenophene > thiophene > benzene. The relative reactivities of furan and thiophene can be shown by reaction (5.19).

$$(5.19)$$

Thiophene is very stable and so will react with most electrophilic reagents which will attack benzene. Nitration, sulphonation, acylation and halogenation may be effected by the "usual" reagents. Furan and pyrrole are, as we have seen, very acid-sensitive and non-acidic reagents such as 1-protopyridinium sulphonate (§ 6.2.2, (**28**)) and acetyl nitrate must be used if sulphonation or nitration are to be effected. Friedel–Crafts acylation of furan and pyrrole may be effected without Lewis acid catalysts or with weakly acid catalysts because of their high reactivity, but alkylation, where the product will have an electron-donating substituent and therefore be *very* acid-sensitive, can only be achieved by indirect methods such as Wolff–Kishner reduction of acylated derivatives.

Because five-membered π-excessive aromatic heterocyclic compounds are more reactive than benzene, it is possible to achieve substitution reactions with electrophilic reagents which will not react with benzene. Thus, although furan reacts as a diene towards the dienophile maleic anhydride (see § 5.3), this reagent acts as an electrophile towards pyrrole and the substitution reaction (5.20) is undergone. Pyrrole and thiophene react with aldehydes and ketones as shown in reactions (5.21) and (5.22).

$$(5.20)$$

$$\text{(thiophene)} + H_3C\text{-CO-}CH_3 \xrightarrow{H^+} \text{(bis-thienyl compound with } H_3C, CH_3\text{)} \tag{5.21}$$

$$\text{(pyrrole)} + H_3C\text{-CO-}CH_3 \xrightarrow{H^+} \text{(octamethylporphyrinogen macrocycle)} \tag{5.22}$$

The "normal" mechanism for electrophilic substitution (cf. (5.17) and (5.18)) holds for substitution of thiophene and many of the reactions of pyrrole and furan. There is evidence, however, that, especially with furan, substitution may be effected by an addition-elimination mechanism as in (5.23). Displacement reactions such as (5.24) are a more common alternative to substitution than is the case with benzenoid aromatics.

$$\text{(furan)} \xrightarrow{CH_3CO_2^{\ominus}NO_2^{\oplus}} \text{(}CH_3COO\text{-O-}NO_2\text{ adduct)} \longrightarrow \text{(}O\text{-}NO_2\text{ furan)} \tag{5.23}$$

$$H_3C\text{-(furan)-}CO_2H \xrightarrow{NO_2^{\oplus}} H_3C\text{-(furan)-}NO_2 + CO_2 \tag{5.24}$$

§ 5.4.1   Substituent effects in electrophilic substitution

As we have seen (§ 5.4), the heteroatom of π-excessive aromatic compounds is responsible for directing electrophilic aromatic substitution to the α- rather than to the β-positions of the ring. The ratio of α-substitution to β-substitution depends both on the heteroatom and on the reagent used. Substituents will have their own directing effect on substitution, and the final positions of substitution will depend on the balance between the effects of the heteroatom and of the substituent.

The effect of an electron-withdrawing group (*meta* directing in benzene) at C_3 is to reinforce the α-directing effect of the heteroatom in directing substitution to C_5 as in reaction (5.25). An electron-donating group at C_2 (*ortho*:*para* directing in benzene) also supports the α-directing effect in directing substitution to C_5 although some 3-substitution ("*ortho*" to the substituent) may also be observed (cf. reaction (5.26)).

$$\text{(3-CHO furan)} \xrightarrow{NO_2^+} \text{(}O_2N\text{-5, 3-CHO furan)} \tag{5.25}$$

$$\text{(2-CH}_3\text{ thiophene)} \xrightarrow{NO_2^+} O_2N\text{-(thiophene)-}CH_3 + \text{(3-NO}_2\text{ thiophene)-}CH_3 \tag{5.26}$$
$$\text{(70 per cent)}$$

An electron-donating group at C_3 will reinforce the α-directing effect in causing substitution to occur at C_2 as in (5.27), unless the bulk of this substituent blocks the 2-position. In this case, the directing effect of the heteroatom is paramount and 5-substitution ensues as in (5.28). When there is an electron-withdrawing group at C_2 the conflict between the effect of the heteroatom and the substituent causes the position of substitution to vary with the nature of the heteroatom and the reagent, as in reactions (5.29) and (5.30).

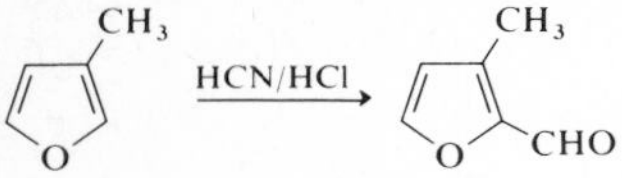

(5.27)

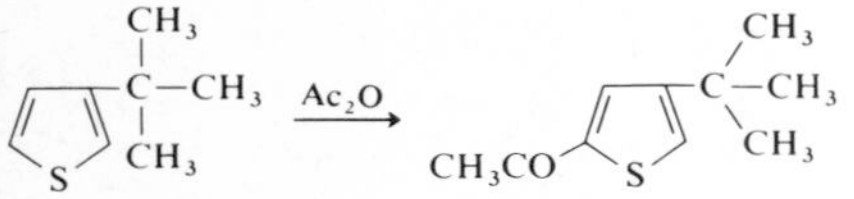

(5.28)

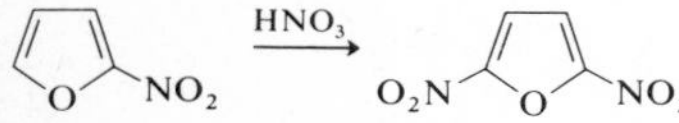

(5.29)

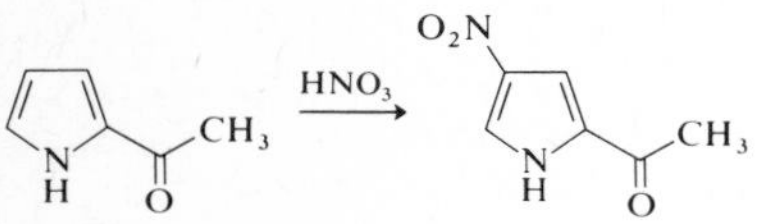

(5.30)

When disubstituted compounds undergo electrophilic aromatic substitution, the position at which attack occurs may be estimated in the same way as for the monosubstituted compounds. Some of the positions for attack are indicated by an arrow in the series of compounds in (5.31).

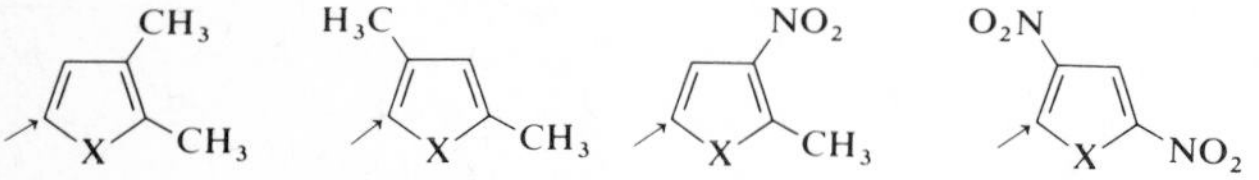

(5.31)

§ 5.4.2   The effect of additional heteroatoms on electrophilic substitution

The effect of introducing an additional nitrogen atom into a five-membered π-excessive compound is similar to that of introducing an electron-withdrawing substituent at that position. Removal of electron density from carbon causes the reactivity towards electrophiles to fall compared to the parent heterocyclic compound, but azoles, the compounds containing one additional nitrogen atom, are still more reactive towards electrophilic reagents than benzene. Two additional nitrogen atoms cause susceptibility to electrophilic attack to fall off further and triazoles (37), oxadiazoles (*e.g.* (38)) and thiadiazoles (*e.g.* (39)) are resistant to electrophilic attack unless a very powerful electron-donating substituent is present.

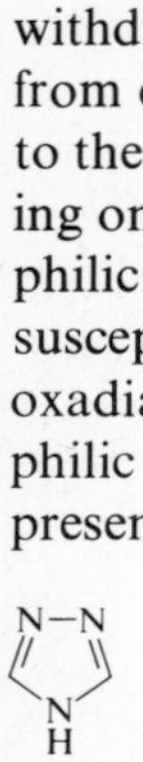

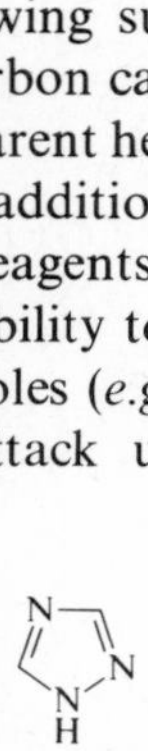

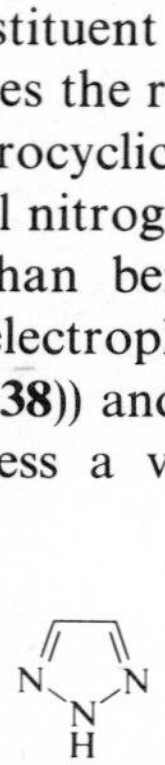

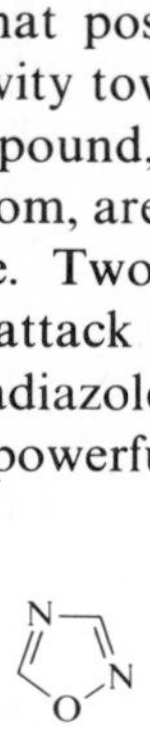

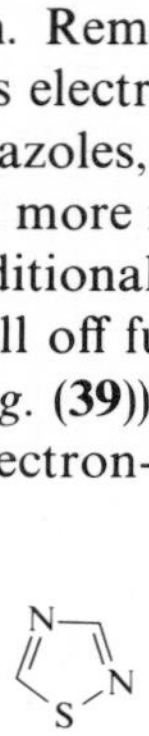

The directing effect of the additional heteroatom, like that of an electron-withdrawing substituent, is what would be termed *"meta"* in benzene chemistry. The combination of the *"meta"*-directing effect of the "doubly bound" nitrogen and the α-directing effect of the "singly bound" heteroatom is to support substitution at C_5 in 1,3-azoles such as imidazole (14), oxazole (15) and thiazole (16). The two effects are contradictory in the 1,2-azoles, pyrazole (11), isoxazole (12) and isothiazole (13). The position of electrophilic attack can, however, often be deduced by consideration of the various canonical forms of a compound. Thus, as we have seen in § 5.1, the resonance forms (17c) and (17d) direct attack to C_4 or C_5 in 1,3-azoles and (18e) directs attack to C_4 in 1,2-azoles, in neutral conditions.

(11)   (12)   (13)   (14)   (15)   (16)   (17c)   (17d)   (18e)

In the acidic conditions of nitration and sulphonation, the species attacked is not the neutral molecule but the cation. Attack will be less readily undergone in acid than in neutral media since the electrophile must now attack an electron-deficient species. The direction of attack in acidic media may be deduced from the most likely Wheland intermediate. Since intermediate (40) has greater charge separation than intermediate (41), substitution at C_5 will be preferred to substitution at C_4.

(40)

or   → Products   (5.32)

(41)

1,2-Azoles are much less reactive than 1,3-azoles and do not usually undergo electrophilic substitution in acid conditions. Since the order of reactivity towards electrophiles in the parent hydrocarbons is N > O > S, the order in the azoles is:

imidazole > oxazole > thiazole; pyrazole > isoxazole > isothiazole

In basic conditions, 1,3-azoles undergo electrophilic substitution reactions, such as deuteration, at C_2. Here the intermediate anion (42) is stabilised by (a) sp^2 hybridisation (increased s character stabilises anionic character); and (b) flanking electron-withdrawing heteroatoms (HC≡N is more acidic than HC≡CH). Electron-withdrawing substituents or additional heteroatoms increase the rate of exchange. Thiazole may owe some of its acidity to the empty d orbitals on sulphur which can overlap with the σ orbital of the anion and cause stabilisation. Both thiazole and isothiazole undergo base-catalysed deuterium exchange at positions adjacent to the sulphur atom, C_2 and C_5 respectively. Isoxazole anions are unstable and undergo ring fission reactions (cf. § 5.9).

(42)

§ 5.5   Quaternary salts and N-acyl substituted compounds

In § 5.2.2 we have seen that, while the monoheteroatomic π-excessive heterocyclic compounds do not readily form quaternary salts, 1,2- and 1,3-azoles do. The resultant positive charge makes for a more acidic proton at C_2 in 1,3-azolium salts due to the increased electronegativity of the heteroatom, and so base-catalysed deuteration is more readily undergone than with unquaternised compounds. The rate of exchange is dependent on the electronegativity of the heteroatom and follows the order oxazolium > thiazolium > imidazolium. 1,2-Azolium salts exchange at C_3 and C_5 in base but more slowly than the 1,3-azolium salts.

The action of thiamine as a coenzyme (see Chapter 8) has stimulated much research into the reactivity of thiazolium salts in base. In general, the salts are affected by extra heteroatoms and electron-withdrawing groups in much the same way as are the free bases but they are more reactive due to the increased electronegativity of the positively charged heteroatom. Ylide stability plays a part in the ease of formation of inter-mediates such as (43).

(43)   (a)   (44)   (b)

We have noted in § 5.2.2 that pyrrole and its aza-analogues may be N-acylated. Since the N-acyl derivatives require the lone-pair on nitrogen for aromatic stability, amide resonance of the usual type (44a ↔ 44b) is inhibited. This is reflected in an atypical carbonyl band in the infra-red spectrum and in enhanced reactivity of the carbonyl group towards nucleophilic reagents. Additional heteroatoms with their electron-withdrawing effect on the "amide" lone pair cause enhancement of this reactivity. N-Acyl tetrazoles are very much more readily hydrolysed than N-acyl pyrroles. Use is made of the high reactivity of N-acyl-imidazoles such as (45) in peptide synthesis as in (5.33) below, where X and Y are protecting groups. N-Acyl imidazoles may be reduced by lithium aluminium hydride to aldehydes as in reaction (5.34) for much the same reasons as N-acylaziridines are reduced to aldehydes (§ 3.2).

$$XNH-\underset{\underset{H}{|}}{\overset{\overset{R}{|}}{C}}-CO_2H \xrightarrow{\quad\quad\quad\quad} XNH-\underset{\underset{H}{|}}{\overset{\overset{R}{|}}{C}}-CON$$

(45)

$$H_2N-\underset{\underset{H}{|}}{\overset{\overset{R'}{|}}{C}}-CO_2Y$$

$$XNH-\underset{\underset{H}{|}}{\overset{\overset{R}{|}}{C}}-CONH-\underset{\underset{H}{|}}{\overset{\overset{R'}{|}}{C}}-CO_2Y \qquad (5.33)$$

$$RCO-N \xrightarrow{\ LiAlH_4\ } RCHO + \qquad\qquad (5.34)$$

§ 5.6   Reactions of substituent groups

§ 5.6.1   Benzene-like behaviour

One feature of aromatic reactivity in benzenoid compounds is the applicability of the Hammett relationship (5.35) to the rates of reactions of substituent groups.

$$\text{Log}\frac{k}{k_0} = \rho\sigma \qquad\qquad (5.35)$$

In this equation, k and k_0 are the rate constants for the reaction of the substituted and unsubstituted compounds respectively, ρ is a parameter characteristic of the reaction under study and σ is a parameter characteristic of the substituent and represents the ability of the group to attract or repel electrons.

Hammett and others have made attempts to apply equation (5.35) to five-membered π-excessive heterocyclic aromatic compounds by considering that since the heteroatom replaces a C=C bond of benzene, then groups related as in (46) may be regarded as *meta* and groups related as in (47) are "*para*". By using values of σ_{meta} and σ_{para} found for benzenoid compounds, straight line plots of $\log(k/k_0)$ versus σ were found for various reactions. The reaction parameter ρ was not however the same as it had been in the corresponding reaction of the benzenoid compound. The ratio $\rho_{\text{HETEROCYCLE}}/\rho_{\text{BENZENE}}$ is nearest unity for thiophenes and the ability of the heteroatom to transmit electronic effects has been found from work on polarographic reduction of nitro compounds to follow the order:

$$HC{=}CH \doteqdot S < O < NH \qquad\qquad (5.36)$$

(46)   (47)

5-Substituted 2- and 3-methyl thiophenes have nmr chemical shift values for methyl groups which give straight line plots against the σ value of the substituent. Rates of hydrolysis of substituted thiophene carboxylic esters and pK_a values of thiophene carboxylic acids can be subjected fairly successfully to the Hammett treatment.

§ 5.6.2   *Reactions not typical of substituted benzenes*

From (5.36) above, it is not surprising that the pK_a of thiophene-3-carboxylic acid is similar to that of benzoic acid. Pyrrole-3-carboxylic acid is, however, a weaker carboxylic acid since the vinylogous urethane character in (48) stabilises the undissociated form of the acid. Pyrrole-2-aldehyde (49) has vinylogous amide character and so, unlike the corresponding furan and thiophene aldehydes, it does not undergo reactions typical of aldehydes and shows carbonyl absorption in the amide region of the infra-red spectrum.

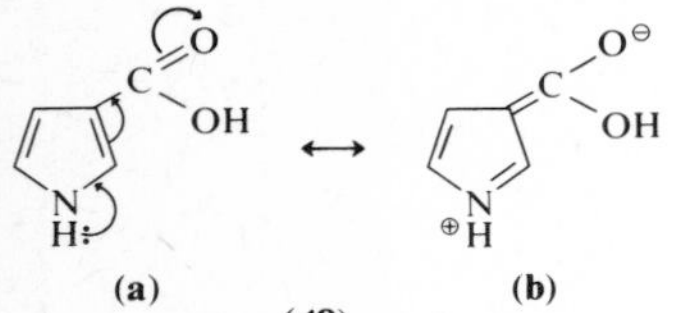

(a)   (b)

(48)

(a)   (b)

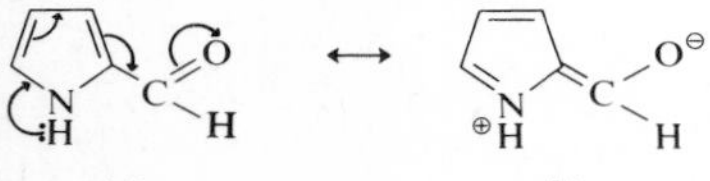

(49)

§ 5.6.3   *Reactions of alkyl substituents*

Unlike 2- and 4-alkyl pyridines (§ 6.6), the alkyl derivatives of π-excessive heterocyclic compounds have no special activation at the side chain and the reactivity is akin to that of alkyl-benzenes. In acid conditions, aldehydes will react as electrophiles at the α-position of the ring rather than at the side chain, as shown in reaction (5.37). Bromination under normal conditions occurs first at ring carbon atoms but N-bromosuccinimide may react with 2-methylthiophene to give side-chain bromination. In the azoles, the doubly bound nitrogen is analogous to the nitrogen in pyridine and so methyl groups α to this heteroatom may react in the same way as do α- and γ-methyl-pyridines (§ 6.6). This is especially marked if the nitrogen is made more electronegative by protonation in acidic media as in (5.38) or by quaternisation as in (5.39).

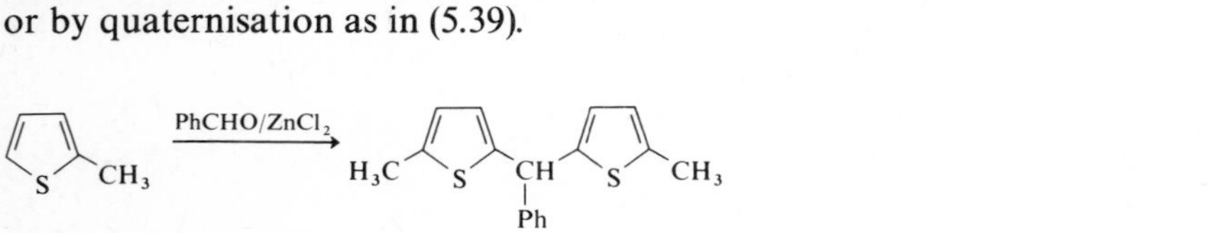

(5.37)

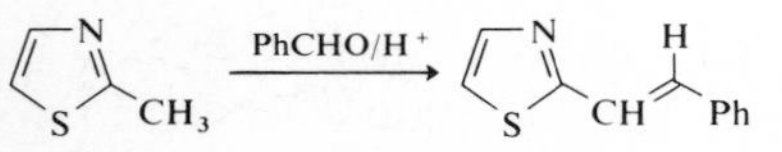

(5.38)

(5.39)

§ 5.6.4   Hydroxymethylene substituents

These have a very labile hydroxyl group and acid-catalysed polymerisation via carbonium ions is observed with furan and thiophene as in reaction (5.40). The pyrrole derivative can act as an electrophile and will substitute on pyrrole as in reaction (5.41).

(5.40)

(5.41)

§ 5.6.5   Halogen and metal substituents

Halogenated compounds can be prepared by direct halogenation and, like most aryl halides, they are comparatively unreactive. 2-Chloro and 2-iodofurans can be converted to Grignard derivatives and 3-iodofuran will yield a lithium derivative. Mercury salts can react directly with furan to yield 2-mercury derivatives. 2-Chlorothiophene does not form Grignard reagents but it can be converted to the sodium salt. Direct lithiation and mercuration can be achieved with butyl lithium and mercury salts respectively. The halogen and metal derivatives react much like their benzenoid counterparts and are very useful in synthesis.

§ 5.6.6   Amino, hydroxy and mercapto substituents

The chemistry of five-membered heterocycles with such substituents is dealt with in Chapter 4.

§ 5.7   Nucleophilic substitution reactions

Phenyl halides are very unreactive towards nucleophilic reagents and π-excessive heteroaromatic halides are equally unreactive. The presence of an electron-withdrawing group in an "*ortho*" or "*para*" relationship to the halogen substituent makes nucleophilic substitution more likely both with benzenoid and with five-membered heteroaromatic compounds, and, whilst substitution reactions are unlikely with 2-bromothiophene, 5-nitro-2-bromothiophene does react with nucleophilic reagents as in (5.42) below. Halogen-substituted azoles with their pyridine-like nitrogen atom should undergo nucleophilic substitution reactions in much the same way as the corresponding pyridine compounds (see § 6.5). The juxtaposition of the halogen and the C=N group should be similar to the corresponding pyridines and so 1,2-azoles substitute readily if halogenated at C_3 and C_5 as in reaction (5.43), while 1,3-azoles substitute if halogenated at C_2 as in reaction (5.44). Fairly mild nucleophilic reagents may be used to achieve these substitution reactions.

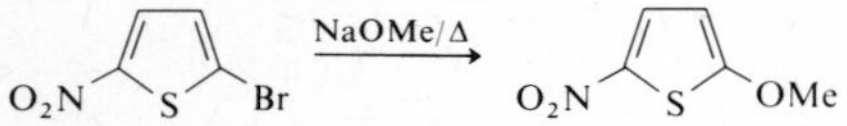

(5.42)

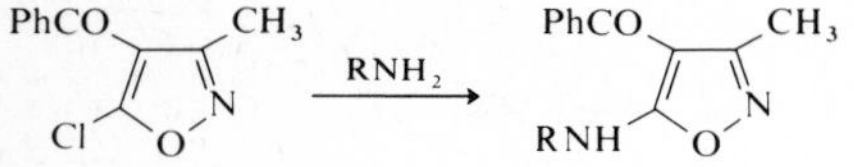

(5.43)

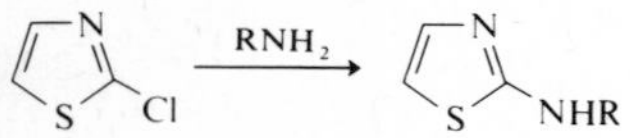

(5.44)

§ 5.8 Radical and carbene reactions

Furan reacts with free radicals preferentially in the 2-position as in (5.45), while the azoles seem less discriminate in their reactions with radicals. Carbenes generated from diazo compounds react with furan, and thiophene and with pyrroles which are substituted at nitrogen with an electron-withdrawing group as in reaction (5.46). Pyrrole itself reacts with chloroform and base (a source of dichlorocarbene) to yield either the aldehyde as in (5.48) or the pyridine as in (5.47) via the adduct (**50**).

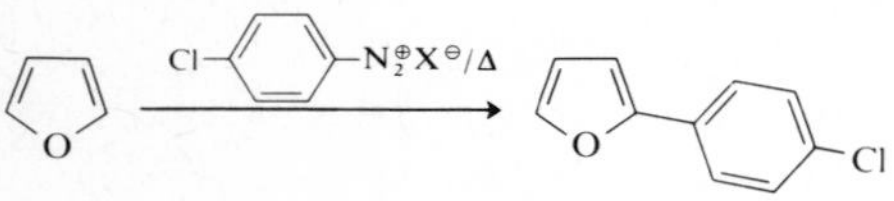

(5.45)

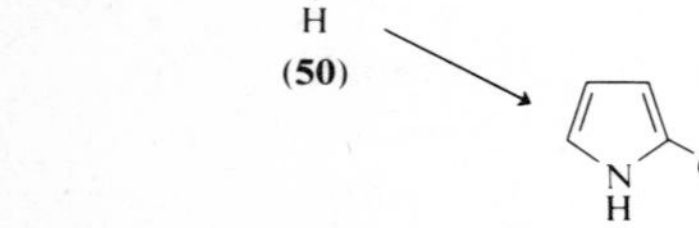

(5.46)

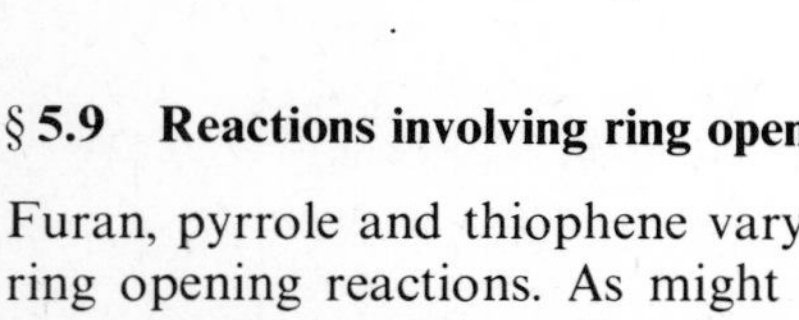

(5.47)

(**50**)

(5.48)

§ 5.9 Reactions involving ring opening

Furan, pyrrole and thiophene vary in the ease with which they undergo ring opening reactions. As might be expected, the least aromatic ring, furan, can be opened readily with a variety of reagents and this is exemplified by the acid catalysed hydrolysis (5.49). Pyrrole is less readily cleaved but may yield a dioxime on refluxing in alcoholic hydroxylamine as in (5.50). Thiophene is very stable but Raney–Nickel desulphurisation (5.51) will cause reductive ring opening.

$$\text{R—[furan]—R} \xrightarrow{\text{H}^+/\text{H}_2\text{O}} \text{R—CO—CH}_2\text{—CH}_2\text{—CO—R} \tag{5.49}$$

$$\text{[pyrrole, NH]} \xrightarrow[\text{ROH}]{\text{H}_2\text{NOH.HCl}} \left[\text{H—CO—CH}_2\text{—CH}_2\text{—CO—H} \right] \rightarrow \underset{\text{NOH}}{\text{CH}} \quad \underset{\text{NOH}}{\text{CH}} \tag{5.50}$$

$$\text{R—[thiophene, S]—R} \xrightarrow{\text{Raney Ni}} \text{R—[chain]—R} \tag{5.51}$$

In the azoles, the oxygen compounds are again most prone to ring-opening reactions. The 1,3-azole, oxazole readily undergoes the acid-catalysed ring-opening reaction (5.52) and the 1,2-azole, isoxazole, undergoes rupture to a β-keto-nitrile in base as in reaction (5.53). Decarboxylation of 3-carboxy-isoxazoles is also accompanied by rupture of the ring as in (5.54) and this reaction has been used in synthesis.

$$\text{R—[oxazole]—R'} \xrightarrow{\text{H}^+/\text{H}_2\text{O}} \text{R—CO—CO—NH—R'} \tag{5.52}$$

$$\text{R—[isoxazole, H, B}^\ominus\text{]} \rightarrow \text{R—CO—CH}_2\text{—CN} \tag{5.53}$$

$$\text{R—[isoxazole, C=O, O}^\ominus\text{]} \rightarrow \text{R—CO—CH}_2\text{—CN} \tag{5.54}$$

Quaternisation of azoles makes for very ready anion formation (see § 5.5), and, while isothiazolium and pyrazolium salts undergo reversible proton exchange, the unstable isoxazolium anion opens in mildly basic conditions to yield the unstable α-ketoketimine as in (5.55). This ketimine reacts readily with acids to yield activated esters and is thus useful in peptide synthesis.

$$\text{R—[isoxazolium, H, B}^\ominus\text{, N—CH}_3\text{]} \rightarrow \left[\text{R—CO—CH=C=NMe} \right] \tag{5.55}$$

§ 5.10 Mesoionic compounds

Mesoionic compounds are heterocyclic compounds which have some degree of aromatic stabilisation but for which no structure can be written using classical structural theory. The first compounds which could be termed mesoionic were the sydnones (first made in Sydney, Australia), for which resonance forms (**51a**) and (**51b**) are obviously five-membered 6π systems. The structure may be written as in (**52**) and nmr spectra and dipole moment studies are in accord with the aromatic dipolar structure.

Electrophilic aromatic substitution reactions such as halogenation, nitration and sulphonation can occur at C_4.

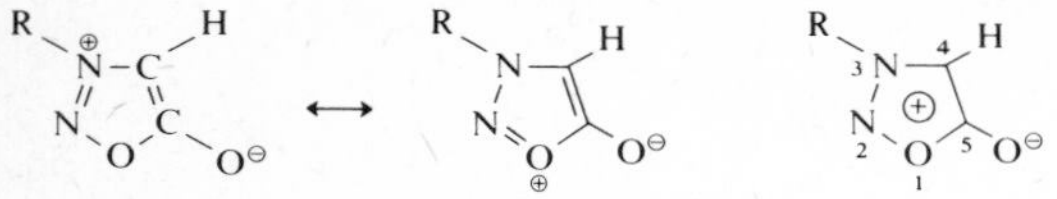

(51a)   (51b)   (52)

More recently a group at Munich have made the mesoionic oxazolone (53) and, in their studies of this compound and of sydnones, they have found that typical aliphatic reactions such as the 1,3-dipolar additions (5.56) and (5.57) were very readily undergone and that these reactions could be used synthetically to make pyrazoles and pyrroles.

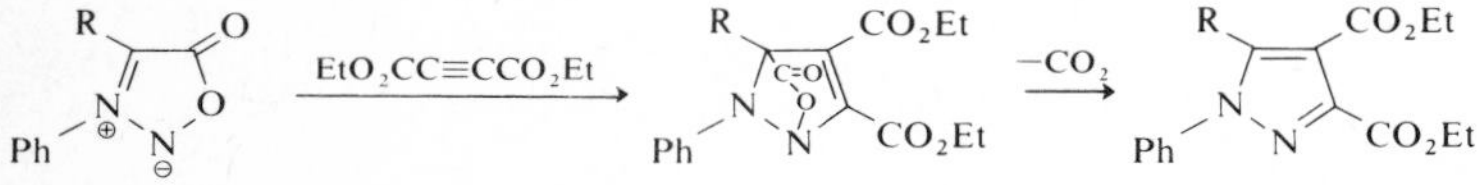

$$\text{EtO}_2\text{CC} \equiv \text{CCO}_2\text{Et} \qquad\qquad -CO_2 \tag{5.56}$$

$$\text{EtO}_2\text{CC} \equiv \text{CCO}_2\text{Et} \qquad\qquad + CO_2 \tag{5.57}$$

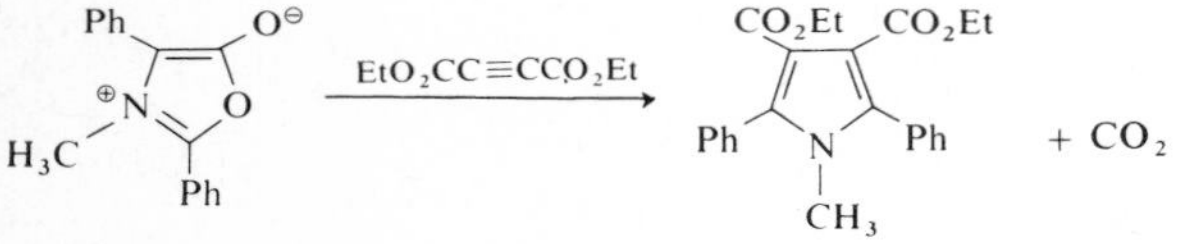

(53)

A variety of other mesoionic compounds such as (54) and (55) have been made and the concept has been extended to larger systems than the simple five-membered heteromonocyclic compounds.

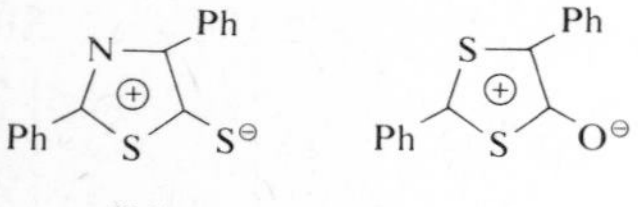

(54)   (55)

§ 5.11   Synthesis of five-membered aromatic heterocyclic compounds

§ 5.11.1   General methods

A useful method for synthesis of a large number of five-membered aromatic heterocyclic compounds is reaction of 1,4-dicarbonyl compounds (e.g. (56)) with a variety of reagents. Thus reaction (5.58) will yield furans, reaction (5.59) pyrroles and reaction (5.60) thiophenes. This method has been termed *Paal–Knorr* synthesis and 1,3-azoles may be synthesised from the dicarbonyl compound (57) as shown in reactions (5.61), (5.62) and (5.63). The main limitation on this method is the availability of the dicarbonyl compound.

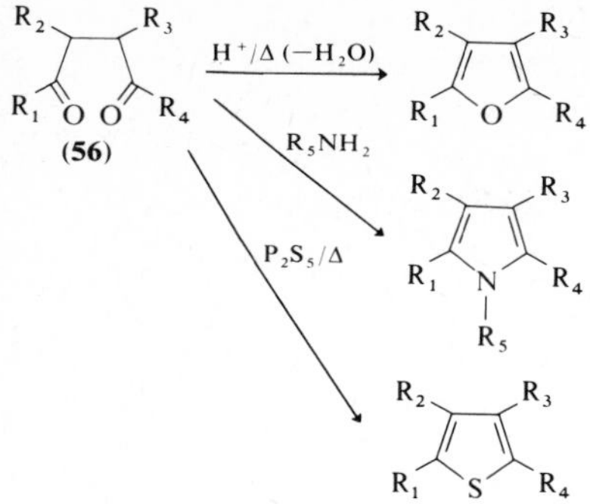

(5.58)

(5.59)

(5.60)

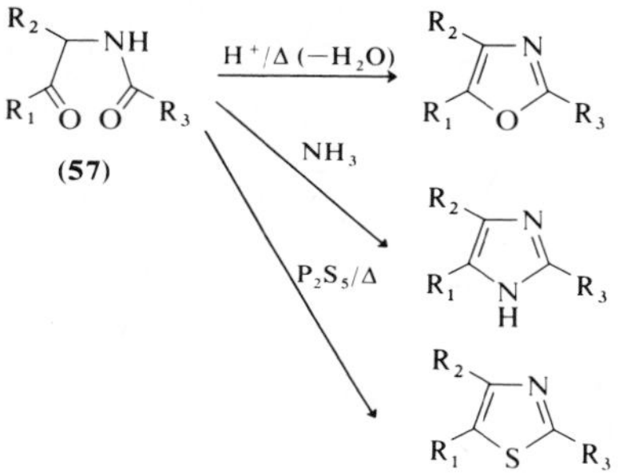

(5.61)

(5.62)

(5.63)

α-Haloketones may be used as starting materials for heterocyclic compounds in this series. In the *Feist–Benary* furan synthesis (5.64), which proceeds by an aldol reaction as the first step, different carbon atoms are joined from the ones which are linked by the *Hantzsch* pyrrole synthesis (5.65). Amidines, thioamides and amides react with α-haloketones to yield imidazoles (5.66), thiazoles (5.67) and oxazoles (5.68) respectively.

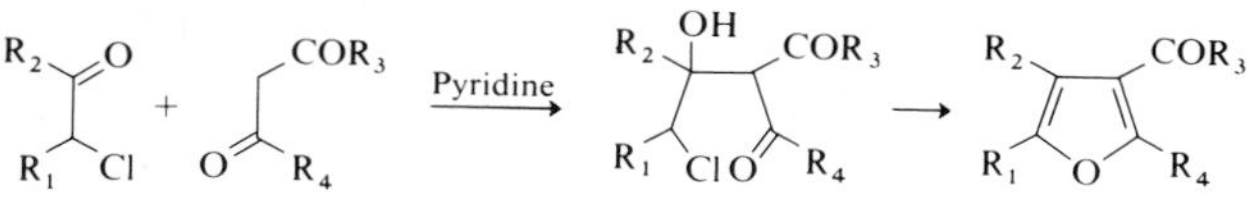

(5.64)

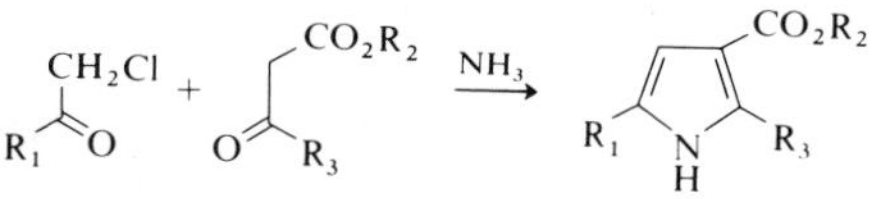

(5.65)

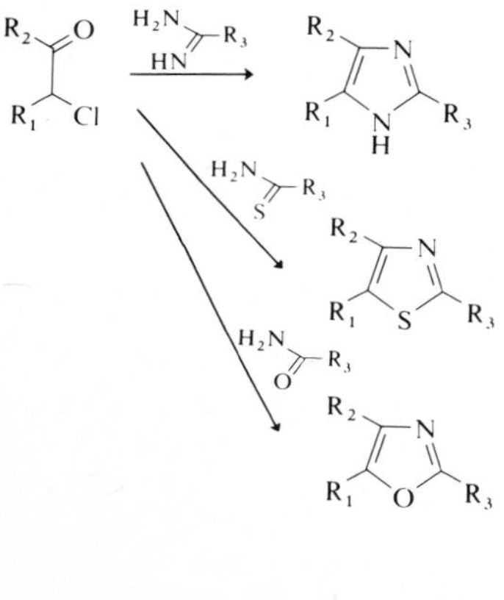

(5.66)

(5.67)

(5.68)

1,3-Dipolar addition, discussed in Chapter 2 (§ 2.8), is a method which has been used to synthesise a large variety of five-membered aromatic heterocyclic compounds. Usually, addition of a 1,3-dipolar compound such as a nitrile oxide, a diazo compound or an azide to a triple bond will yield a five-membered heteroaromatic compound, and oxazoles, pyrazoles, triazoles and tetrazoles have been made in such a way as shown in (5.69), (5.70), (5.71) and (5.72) respectively, below.

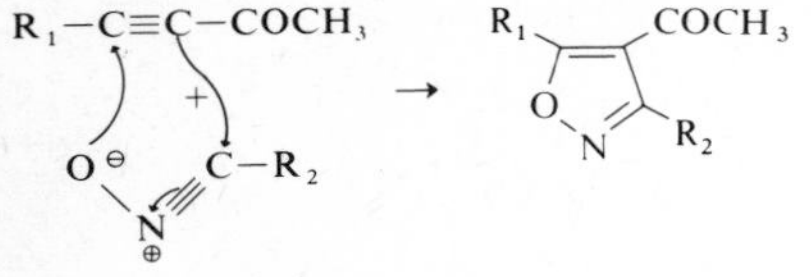

(5.69)

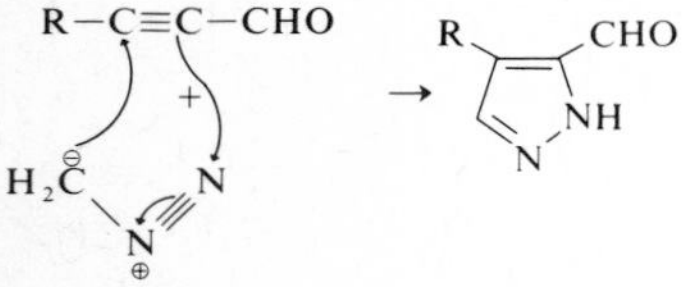

(5.70)

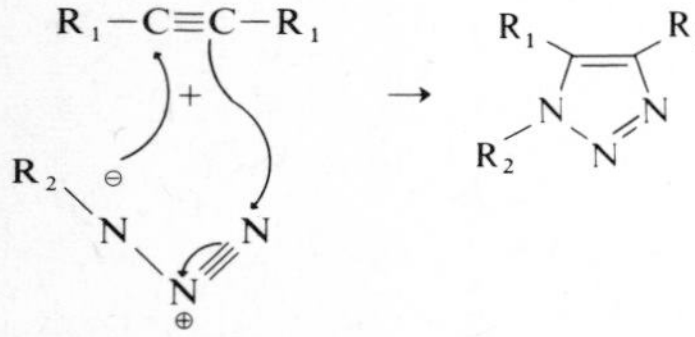

(5.71)

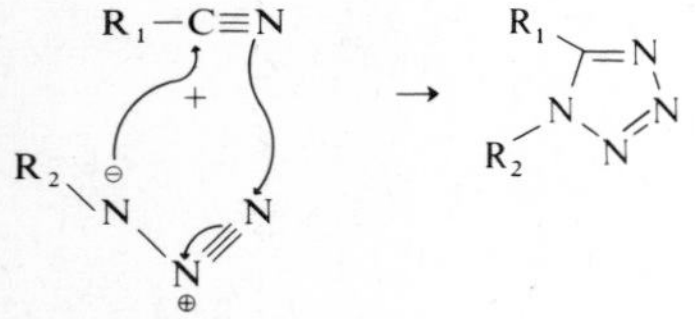

(5.72)

§ 5.11.2 Specific methods

(1) *Furan* is obtained commercially as the derivative furfural by acid catalysed dehydration of waste oat husks which contain xylose, as in (5.73).

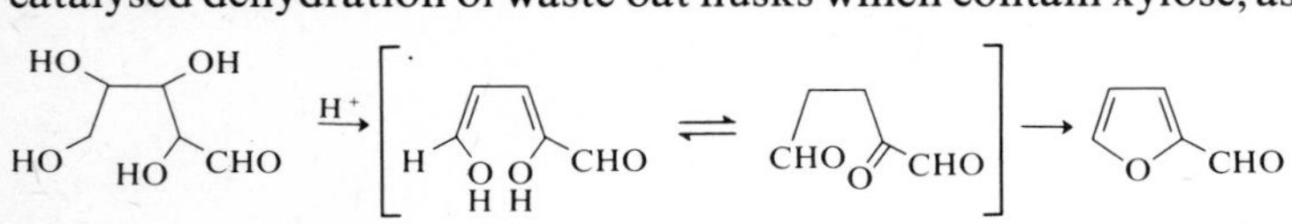

(5.73)

(2) *Pyrroles* are generally synthesised by the *Knorr* synthesis which utilises the reaction of an α-amino carbonyl compound with a ketone or ester. Since α-amino ketones are prone to dimerise they are usually generated *in situ* by reduction of α-oximino ketones which themselves are easily made from ketones. An example of the Knorr synthesis is given in (5.74) below.

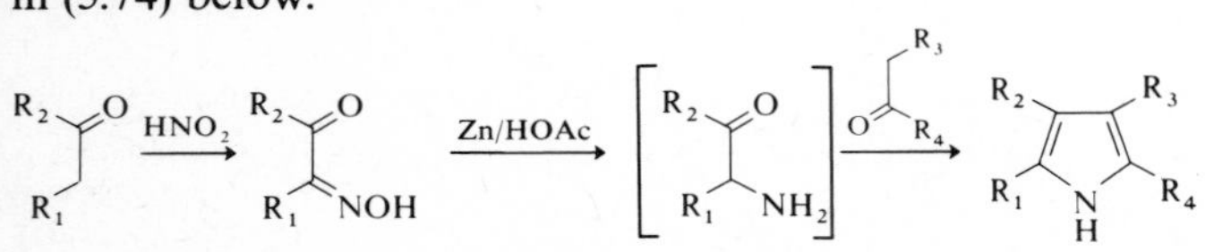

(5.74)

(3) *Thiophenes* may be made by the *Hinsberg* synthesis using α-diketones and diethylthiodiacetate as in (5.75) or from acetylenic compounds as in (5.76).

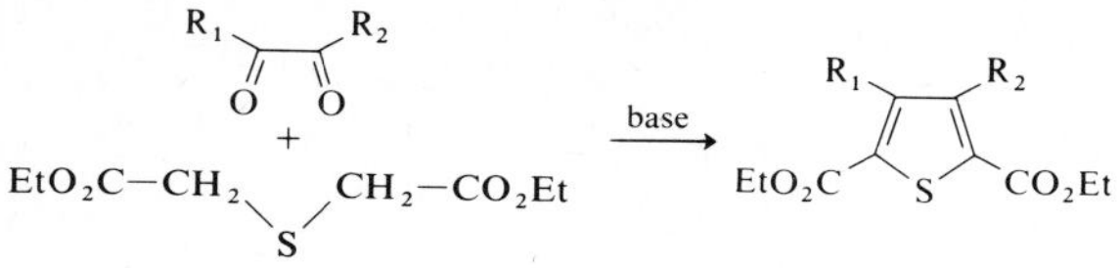

$$(5.75)$$

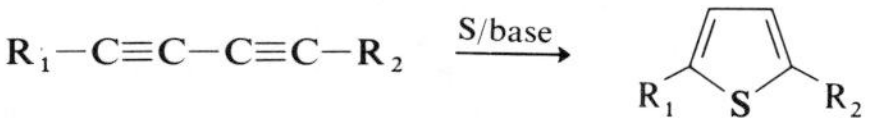

$$(5.76)$$

(4) 1,2-*Azoles*, such as pyrazoles or isoxazoles, may be made by treatment of 1,3-diketones with hydrazine or hydroxylamine as in (5.77) and (5.78) below. When unsymmetrically substituted diketones are used, mixtures may result.

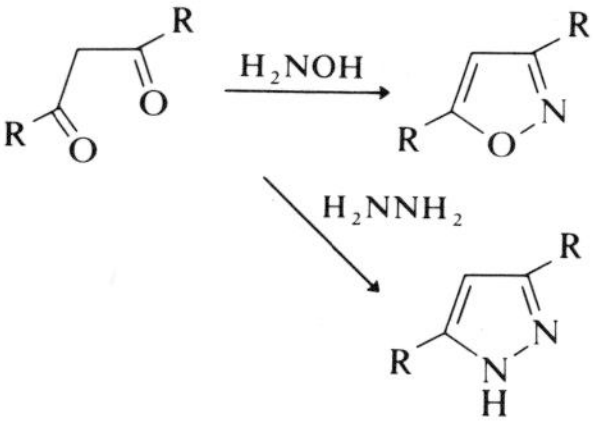

$$(5.77)$$

$$(5.78)$$

Key points

(1) Five-membered heterocyclic compounds of general formula (**1**) can have aromatic properties due to delocalisation of a 6π-electron system which includes a lone-pair from the heteroatom and the four "olefinic" electrons.

(2) Since resonance requires donation of the heteroatom lone-pair to ring carbons, there is a decrease in the base strength and nucleophilic properties of the heteroatom compared with non-heteroaromatic analogues.

(3) For the same reason there will be a net excess of charge on the ring carbon atoms in such compounds and so the compounds are extremely prone to electrophilic aromatic substitution reactions at carbon.

(4) The electronegativity of the heteroatom in such compounds can, to a large extent, account for the degree of aromatic behaviour or diene-like behaviour in such compounds.

(5) Substituent effects are observed in the chemistry of such compounds.

(6) The presence of additional heteroatoms in such compounds has a predictable effect on their chemistry.

(7) Quaternisation can affect the chemistry of such compounds dramatically.

Further reading

General
R. C. Elderfield, ed., *Heterocyclic Compounds*, vol. 1, Wiley, New York, 1950.

Pyrroles
E. Baltazzi and L. I. Krimen, *Chemical Reviews*, 1963, **63**, 511.
R. A. Jones, *Advances in Heterocyclic Chemistry*, 1970, **11**, 383, Academic Press, New York.

Furan
P. Bosshard and C. H. Eugster, *Advances in Heterocyclic Chemistry*, 1966, **7**, 378, Academic Press, New York.

Thiophene
S. Gronowitz, *Advances in Heterocyclic Chemistry*, 1963, **1**, 1, Academic Press, New York.

Azoles
Imidazole. M. R. Grimmett, *Advances in Heterocyclic Chemistry*, 1970, **12**, 103, Academic Press, New York.
Pyrazole. A. N. Kost and I. I. Grandberg, *Advances in Heterocyclic Chemistry*, 1966, **6**, 347, Academic Press, New York.
Isoxazole. N. K. Kochetkov and S. D. Sokolov, *Advances in Heterocyclic Chemistry*, 1963, **2**, 365, Academic Press, New York.
Isothiazole. R. Slack and K. R. H. Wooldridge, *Advances in Heterocyclic Chemistry*, 1965, **4**, 107, Academic Press, New York.

Electrophilic substitution of five-membered heterocyclic compounds
G. Marino, *Advances in Heterocyclic Chemistry*, 1971, **13**, 235, Academic Press, New York.

Chapter 6

Heterocyclic analogues of benzene

Trivalent nitrogen can replace carbon in double bonds and analogues of benzene such as pyridine (**1**), pyridazine (**2**), pyrimidine (**3**), pyrazine (**4**), 1,3,5-triazine (**5**) and 1,2,4,5-tetrazine (**6**) are known.

(1)   (2)   (3)   (4)   (5)   (6)

Other group-five elements may replace carbon in this way, and phosphorus, arsenic, antimony and bismuth analogues of pyridine ((**7**) X = P, As, Sb, and Bi, respectively) have been prepared. Replacement of carbon by a heteroatom destroys the symmetry of benzene and a study of the chemistry of the group-five compounds (**7**) shows that the aromatic character of the compound falls off as the size of the heteroatom increases. Thus pyridine has no tendency to react as a diene, but the phosphorus analogue undergoes the Diels–Alder reaction (6.1) at temperatures in excess of 100°C. The arsenic compound undergoes this reaction at room temperature and the bismuth analogue is so unstable that it can only be isolated as the adduct (**8**).

$$F_3CC{\equiv}CCF_3$$

(6.1)

(7)   (8)

The boron anion (**9**) is isoelectronic with this series and its metal complexes have been reported to have typically aromatic spectra. Boron may accept electrons in dative bonding from nitrogen, and so it was hoped that borazines (**10**) might be aromatic due to contributions from the canonical form (**11**). There seems to be little evidence to support this. Heteroatoms in the second and higher rows of the periodic table have empty d orbitals and so could expand their shells by dative bonding from nitrogen. There was some hope that phosphonitrilic chlorides such as (**12**) and silicon–nitrogen heterocyclic compounds such as the compound (**13**) would be aromatic but, although there seems to be p–d π-overlap in these

73

compounds, it does not seem to be able to extend to full aromatic conjugation. There is evidence that the compound (13) has a puckered ring.

(9)   (10)   (11)   (12)   (13)

The divalent elements of group six can replace carbon in a benzene ring only if the lone-pair is involved in bonding and the heteroatom has a positive charge. Although pyrilium (14) and thiapyrilium (15) salts are not very stable, they do appear to have some aromatic properties and are much more stable than aliphatic oxonium salts. Since sulphur is a second-row element and can expand its valence shell, thiabenzene (16) can be prepared although the ylide formula (17) seems to be the best representation of the structure of the compound. The apparent lack of any ring current in thiabenzene indicates that, as with phosphonitrilic halides, the d orbitals are non-conducting. As we have seen in Chapter 4, α- and γ-pyrones, pyridones and thiapyrones ((18), and (19), X = O, NH or S) have contributing canonical forms (18b) and (19b) and so may be considered to be partially aromatic.

(14)   (15)   (16)   (17)   (18)   (19)

§ 6.1 Physical properties of six-membered heteroaromatic compounds

§ 6.1.1 Bond lengths and resonance energies

Pyridine has bond lengths intermediate between normal double and single bond lengths (C—C = 1·39 Å and C—N = 1·34 Å) as might be expected of an aromatic compound. Introduction of further heteroatoms makes for less symmetrical structures, and resonance energies fall off as the number of heteroatoms is increased. Resonance energies have been calculated as pyridine (1), 133 kJ mol^{-1}; pyrimidine (3) 110 kJ mol^{-1}; pyrazine (4), 100 kJ mol^{-1}; compared to a value of 150 kJ mol^{-1} for benzene. Introduction of heteroatoms to the benzene structure allows for more canonical forms in the resonance hybrid and the electronegativity of the heteroatom localises negative charge. In pyridine, canonical forms (20a) to (20e) are possible and so positions 2, 4 and 6 in the ring have a partial positive charge. The effect of the heteroatom is very similar to the effect of an electron-withdrawing substituent at that position on a benzene ring.

(20a)   (20b)   (20c)   (20d)   (20e)

The molecular orbital approach furnishes an electron density map of pyridine with an excess of electron density on nitrogen and a deficiency of electron density at carbon atoms 2, 4 and 6.

§ 6.1.2   Dipole moment studies

The dipole moment of 2·2 D for pyridine supports structures with some charge separation. The presence of an electron-donating group at C_4 will increase this charge separation as in (21b) below, and so 4-amino-pyridine has a dipole moment of 4·4 D. Electron-attracting substituents have the reverse effect and 4-cyanopyridine has a dipole moment of 1·6 D. With some substituted pyridines, mesomeric and inductive effects compete and in 4-chloropyridine a dipole moment of 0·78 D would suggest that the inductive effect is dominant.

When there are two nitrogen atoms present in the ring, then each will have an equal affinity for electrons. Where both atoms are at the same side of the ring, as in pyridazine (2), there is a greater pull of electrons to that side and a high dipole moment (4 D) results. When the nitrogen atoms are symmetrically placed, as in pyrazine (4), then the resultant dipole moment is zero.

(21a)   (21b)   (2)   (4)   (22)   (23)

Pyridine will form an N-oxide (see § 6.2.2). Since the dipole moment of pyridine is 2·2 D and that of piperidine-N-oxide (23) is 4·38 D, we might expect a dipole moment of 4·38 + 2·2 = 6·6 D for pyridine-N-oxide. The dipole moment has been shown to be 4·3 D and this low value has been explained as being due to the resonance forms (22e) to (22h). These resonance forms explain many of the reactions of pyridine N-oxide (see § 6.2.2 and § 6.4).

(22a)   (22b)   (22c)   (22d)

(22e)   (22f)   (22g)   (22h)

§6.1.3   NMR spectroscopy

This has, as we have seen (§ 4.2.1), been used to detect aromatic character. The heteroatom makes for an uneven distribution of charge in pyridine and the α-, β- and γ-hydrogens therefore absorb at different chemical shifts. There is, however, evidence of the expected deshielding by the ring current. The nmr spectra of ((7) X = P, As and Sb) have been used as evidence of aromatic character in these compounds. The nmr spectra of pyrilium salts have suggested that these compounds have some degree of aromatic character.

§ 6.2   Reactions involving the heteroatom

Pyridine, diazines, triazines and tetrazines (compounds (1) to (6)) have non-bonding lone-pairs of electrons on nitrogen which allow hydrogen-bonding. This, with the dipole moment, accounts for the water solubility of these compounds.

§6.2.1   Basicity

The fact that the lone-pairs on the heteroatoms in compounds of this series are sp^2 hybrid means that such compounds will be less basic than aliphatic amines which have sp^3 hybrid lone-pairs (see § 2.3). The lone-pairs are not required to maintain the aromatic sextet of electrons as is the case with pyrrole, and, being orthogonal to the aromatic system, cannot conjugate with it. This makes pyridine and similar compounds reasonably basic and stable salts may be formed with strong acids.

Pyridine has a pK_a of 5·2 (aliphatic amine $pK_a \sim 10$) and electron-donating substituents on the ring, by increasing the electron density at the heteroatom, result in an increase in base strength. Conversely, electron-withdrawing groups result in a decrease in base strength. Substituent effects will be more important when the substituent is "*ortho*" or "*para*" (*i.e.* at positions 2, 4 or 6) to the heteroatom since resonance effects will then apply. Substituents at positions 3 or 5 can only act by inductive effects and the substituent effects are therefore less dramatic. In 4-aminopyridine, resonance forms (24a) and (24b) stabilise the salt.

(a)   (b)
(24)   (25)   (26)

Alkyl substituents at positions 2 and 6 can interact sterically with acids. Small substituents (CH_3, CH_3CH_2, etc.) only show this effect as the bulk of the acid used increases from H^+ through BH_3 and BF_3 to $B(CH_3)_3$, and although F-strain (see § 2.3) may make such compounds weaker bases than pyridine towards $B(CH_3)_3$, there is no steric effect towards protic acids. Thus 2,6-lutidine (25) is a stronger base ($pK_a = 6.75$) than pyridine due to the electron-donating effect of the alkyl groups. The very large *tert*-butyl groups in (26), however, do cause base weakening in protic acids ($pK_a = 3.58$).

The effect of adding further nitrogen atoms to a pyridine ring may be likened to the effect of electron-withdrawing substituents and thus diazines are less basic than pyridine (pyridazine (2), $pK_a = 2.33$; pyrimidine (3), $pK_a = 1.3$; pyrazine (4), $pK_a = 0.65$).

§6.2.2   *Other reactions involving the lone pair*
Pyridine forms complexes with metal ions and reacts with alkyl halides and sulphates to give alkyl pyridinium salts as in reaction (6.2). Diazines will react in a similar manner but less readily to form monoalkyl and sometimes dialkyl salts. The position of monoalkylation of diazines is susceptible to substituent effects and in reaction (6.3) the 4-methyl group directs the alkylation as shown.

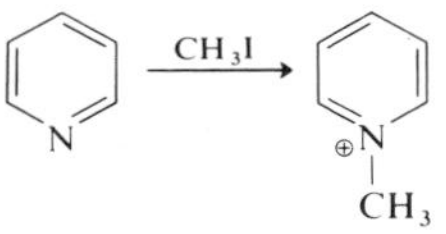

(6.2)

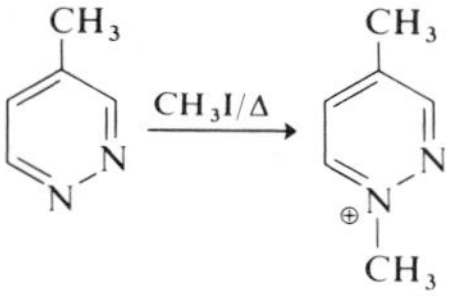

(6.3)

Pyridine is often used as a solvent for the acylation of phenols, alcohols and amines with carboxylic acid chlorides or anhydrides. The actual acylating agent is the very reactive acyl pyridinium salt and the mechanism of acylation is outlined in reaction (6.4). 1-Cyano- and 1-nitropyridinium salts (27) and (29) and the 1-sulphonate (28) are also used as active agents.

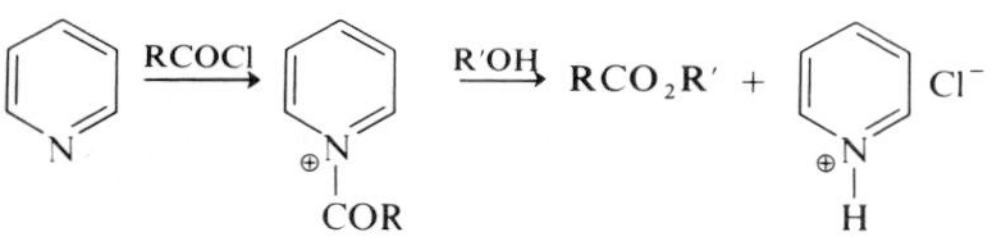

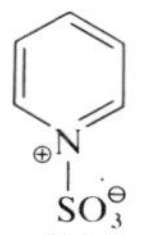

(6.4)

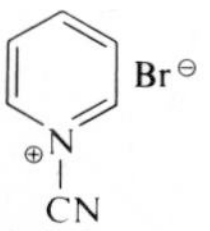

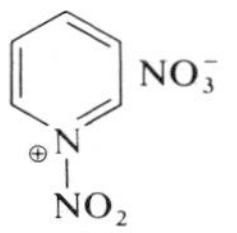

It is interesting that whereas reaction of pyridine or lutidine with SO_3 in liquid sulphur dioxide yields the N-sulphonate such as (28); reaction of 2,6-di*tert*-butylpyridine causes sulphonation at C_3. This must be due to the steric effect of the *tert*-butyl groups protecting the lone-pair.

One of the most important reactions of pyridine and the diazines is the formation of N-oxides as in reaction (6.5). The N-oxides are very important compounds for achieving electrophilic aromatic substitution of pyridines (cf. § 6.4) and by this route synthesis of a variety of substituted pyridines can be achieved. The N-oxides may be reconverted to pyridines as in reactions (6.6) and (6.7).

(6.5)

(6.6)

(6.7)

§ 6.3   Reactions typical of dienes

Such reactions are not common with six-membered aromatic heterocyclic compounds although, as we have seen in the series ((7), X = N, P, As, Sb, Bi), as the size of the heteroatom increases so the tendency to undergo Diels–Alder reactions increases. 2-Pyrone will also undergo Diels–Alder reactions (see § 6.7).

§ 6.4   Electrophilic aromatic substitution reactions

Substitution by electrophilic reagents is a reaction which typifies the chemistry of benzene. Pyridine, with its electronegative heteroatom causing π-deficiency at carbon (resonance forms (20a–e)), will obviously be deactivated towards electrophilic reagents. Such reactions are therefore very slow and many reagents react at nitrogen rather than at carbon, as we have seen in § 6.2.2. When reactions are conducted in acidic medium then pyridinium salt formation will cause further deactivation of the ring. Electrophilic substitution reactions in acidic conditions are therefore rare, nitration only occurring in small yields under forcing conditions and sulphonation occurring only in the presence of catalytic amounts of mercuric sulphate. Halogenation can be achieved, but since aluminium chloride may complex with the ring nitrogen and deactivate the molecule, Friedel–Crafts reactions are unknown in pyridine chemistry. Since in resonance forms (20a–e), the only carbon atoms not to bear a positive charge are those at positions 3 and 5 (the β-positions), electrophilic substitution, when it occurs, will occur at these positions.

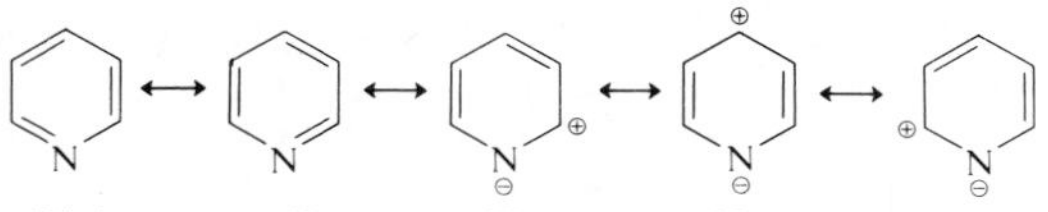

(20a)   (20b)   (20c)   (20d)   (20e)

Electron-donating substituents will make for easier electrophilic aromatic substitution and will direct substitution to the *ortho* or *para* positions as in (6.8) and (6.9). Electron-withdrawing substituents further deactivate the ring to electrophilic aromatic substitution and additional nitrogen atoms in the ring will have the same effect. Electrophilic substitution is therefore extremely rare with diazines unless there are strongly activating substituents present as in reaction (6.10). Triazines are even more resistant to electrophilic aromatic substitution.

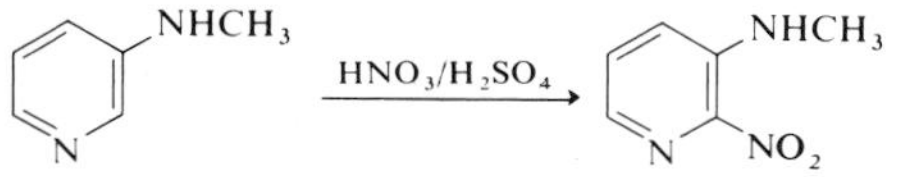

(6.8)

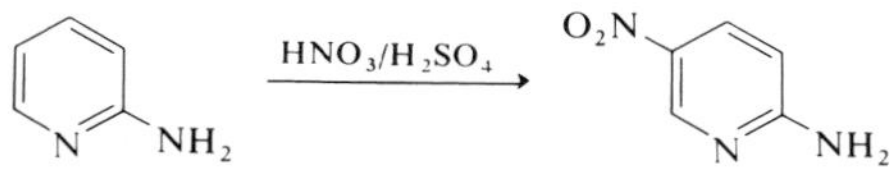

(6.9)

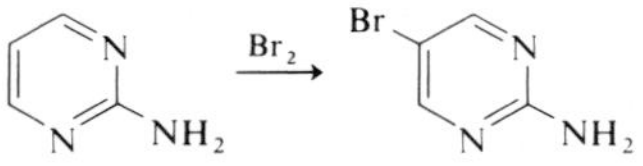

(6.10)

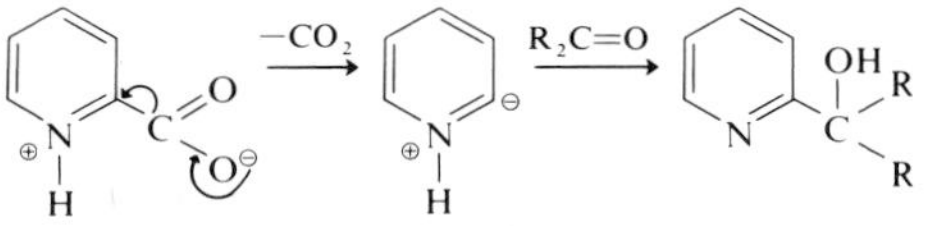

(6.11)

Decarboxylative electrophilic aromatic substitution reactions may be achieved and the zwitterionic picolinic acid can be substituted as in reaction (6.11).

N-oxides. Because resonance forms (22e) to (22h) imply that pyridine-N-oxide will have a net negative charge at positions 2, 4 and 6, the compound is reactive towards electrophilic reagents. Further, because in resonance forms (22e) to (22h) there is a positive charge on nitrogen, repulsive forces make position 4, the furthest from the nitrogen, the most likely of these activated positions to be attacked. In strongly acidic conditions, protonation yielding compound (30) leads to deactivation, and if electrophilic substitution is observed, it occurs at the β-positions (C_3 and C_5).

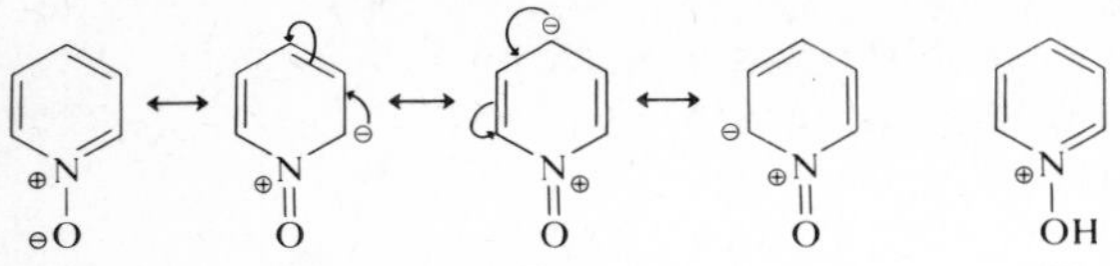

(22e)   (22f)   (22g)   (22h)   (30)

Since the pK$_a$ of pyridine oxide is 0·79, *very* strongly acidic conditions are required for such deactivation, and nitration of pyridine-N-oxide occurs readily in the γ-position as in (6.12). Halogenation occurs at α- and γ-positions and, in strong acid, at the β-position, and sulphonation also occurs at the β-position. Strongly electron-donating substituents may swamp the directing effect of the N-oxide and direct substitution to *ortho* and *para* positions as in (6.13). In pyridine-N-oxide substituted by weaker electron-donating groups, the N-oxide is the directing group as in (6.14).

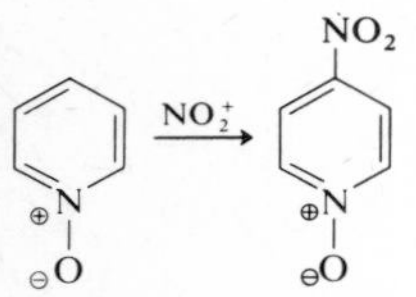

(6.12)

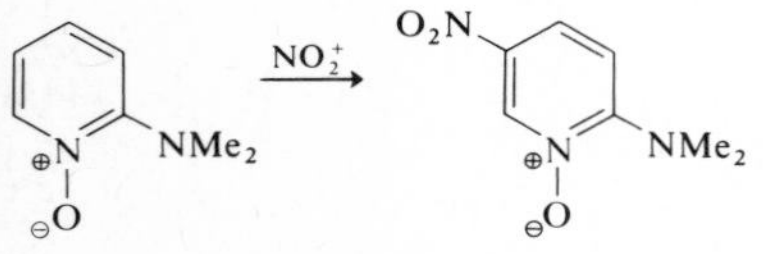

(6.13)

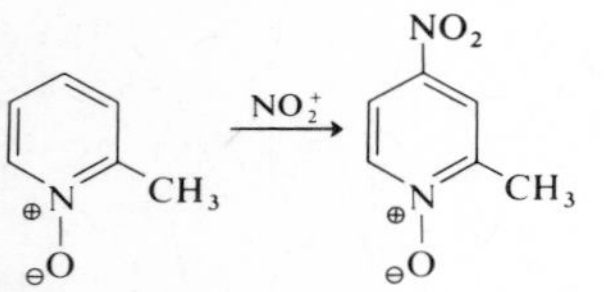

(6.14)

The N-oxides of diazines are also more reactive than the free bases and pyridazine-N-oxide is nitrated at position 4.

N-Oxides may be used synthetically to obtain suitably substituted pyridines and diazines by electrophilic substitution. Removal of the oxygen may then be effected as shown in reactions (6.6) and (6.7) (see § 6.2.2).

§ 6.5 Nucleophilic substitution and addition reactions

Canonical forms (20a) to (20e) imply that pyridine will be deactivated towards electrophilic reagents but, being electron deficient at positions 2, 4 and 6, it will react with strongly nucleophilic reagents at these positions.

Thus sodium amide substitutes pyridine at position 2 (reaction (6.15)) in a mechanism which must involve loss of hydride ion from the intermediate (31). Since hydrogen gas is evolved in the reaction, it is thought that protonation by the product 2-aminopyridine may be involved. This reaction, which was discovered by Tschitschibabin in 1914, always involves substitution at position 2 unless positions 2 and 6 are blocked. In this case the 4-position may be attacked. Other very nucleophilic reagents such as alkyl and aryl lithium can substitute pyridine in the presence of a hydride scavenger such as atmospheric oxygen (reaction (6.16)).

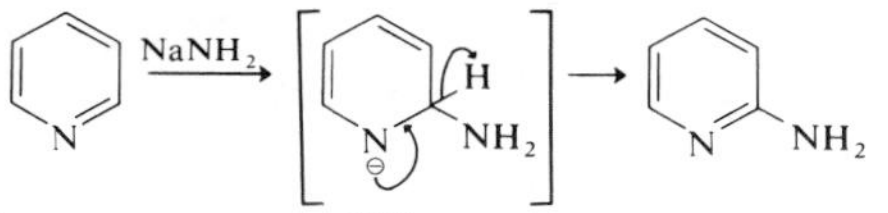

(6.15)

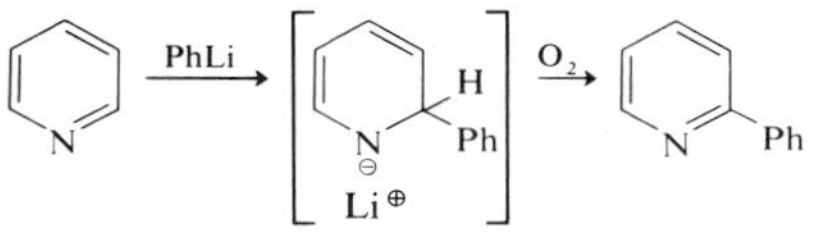

(6.16)

Hydroxide ion, being a much weaker nucleophile, will only substitute at very high temperatures and then in very low yield (reaction (6.17)). Halogenation can be achieved by thionyl chloride but through the intermediacy of a complex such as (32) (reaction (6.18)).

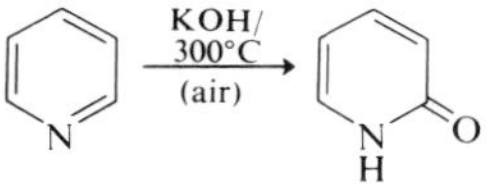

(6.17)

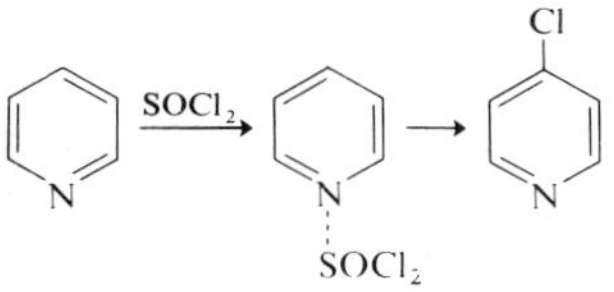

(6.18)

Alkylation of pyridine-N-oxide (reaction (6.7)) yields N-alkoxypyridinium salts, which, because of the positively charged nitrogen, will be more prone to attack by nucleophiles than will pyridine itself. The intermediate adducts eliminate readily to yield the substituted pyridine (reaction (6.19)), but in some cases the nucleophile may react as in (6.20), causing generation of the N-oxide. Pyridinium salts are activated, in a similar way towards nucleophiles although, for the most part, such reactions (e.g. (6.21) and (6.22)) are additions and not substitutions and the 2-position is usually the preferred site of attack.

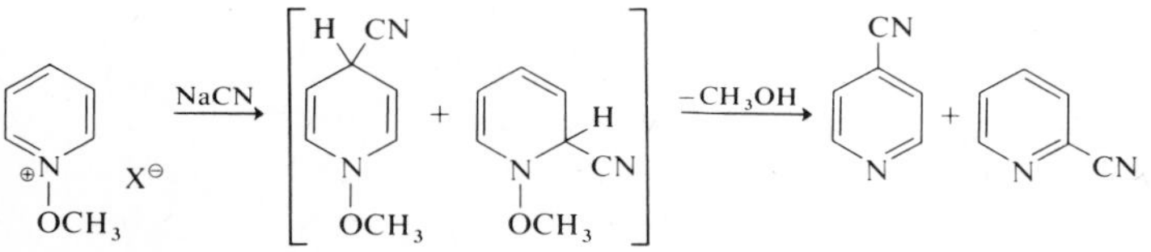

(6.19)

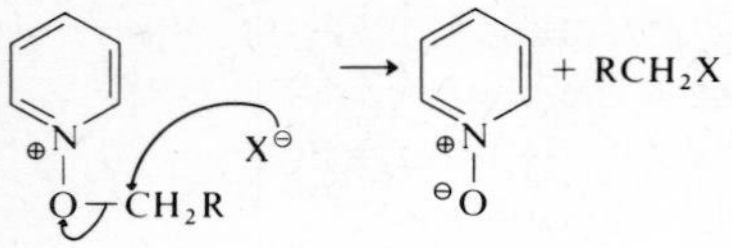

(6.20)

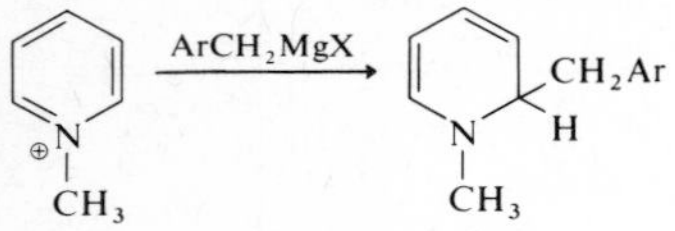

(6.21)

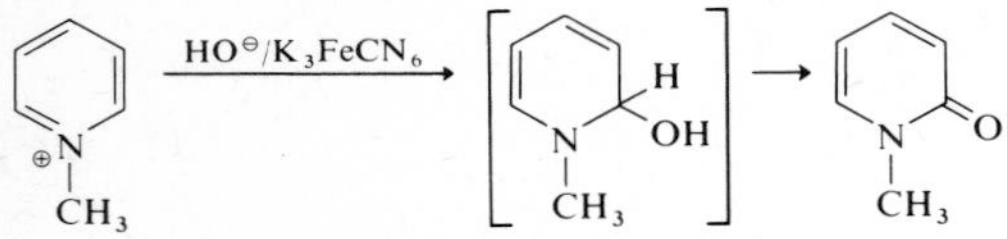

(6.22)

Pyridine, being π-deficient, is more susceptible to *reduction* than benzene and may be reduced by Raney Nickel to piperidine even at atmospheric pressure as in reaction (6.23). Reduction may also be achieved using sodium and alcohol or $LiAlH_4$ and the product is invariably the tetrahydropyridine (cf. (6.24)). Alkylpyridinium salts are more electrophilic than pyridine and so can be reduced even by the mild reductant sodium borohydride. This reagent yields dihydro-derivatives in alkaline conditions as in (6.25). The easy reduction of pyridinium salts is of great importance in biological systems as we shall see in Chapter 8.

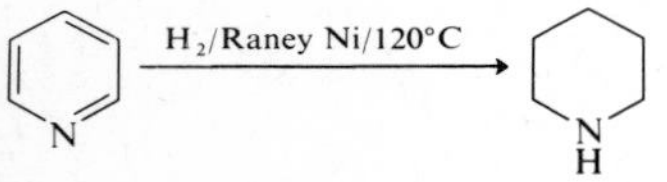

(6.23)

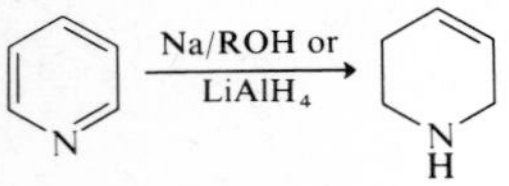

(6.24)

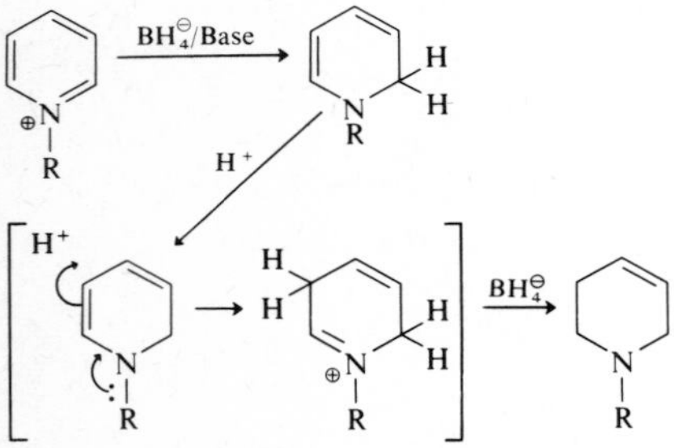

(6.25)

(6.26)

Pyrilium salts are even more reactive towards nucleophiles than are pyridinium salts, but subsequent ring-opening reactions may determine the nature of the product as in reactions (6.27), (6.28) and (6.29). Like pyridinium salts, 2-substitution is preferred. Pyrilium salts can be converted to pyridinium ions and phosphorins as in reactions (6.28) and (6.29) respectively. Thiapyrilium salts are more stable than pyrilium salts and will react with the nucleophilic reagent LiAlH$_4$ at positions 2- and 4- as in reaction (6.30).

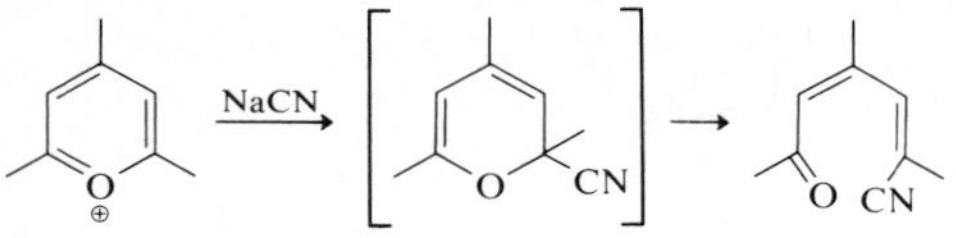

(6.27)

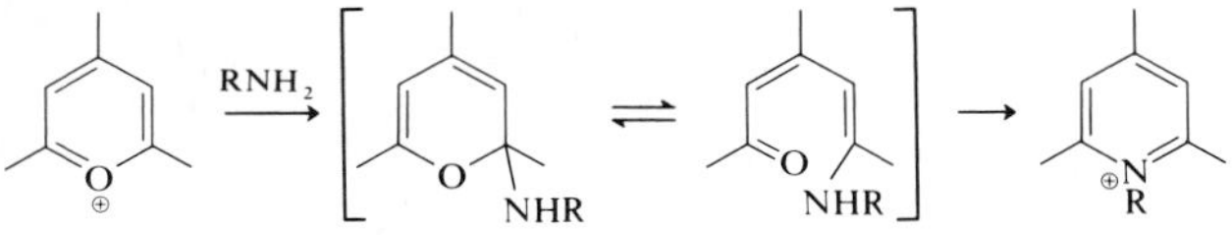

(6.28)

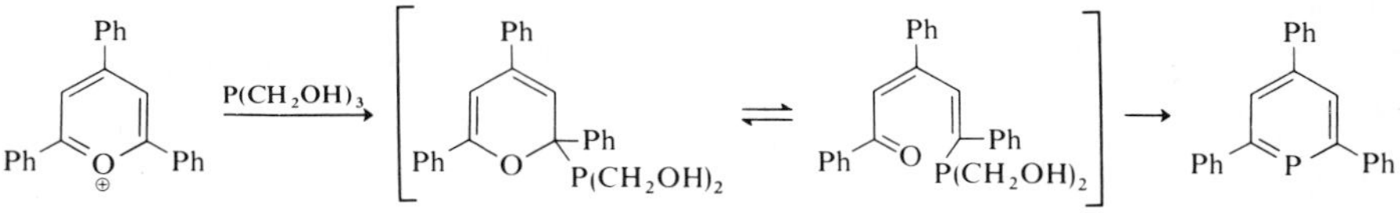

(6.29)

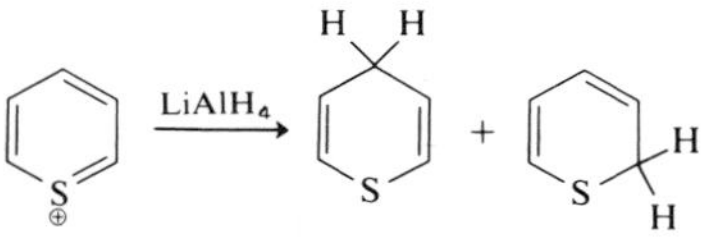

(6.30)

Phenyl halides are very resistant to nucleophilic substitution reactions and indeed nucleophilic substitution is only encountered when there are strong electron-withdrawing groups present. Thus 2,4-dinitrochlorobenzene can react with nucleophilic reagents. 2- and 4-Halogenated pyridines are understandably also prone to nucleophilic substitution reactions (e.g. reactions (6.31) and (6.32)), but 3-halogenated pyridines are more resistant to nucleophilic attack. Quaternisation of the pyridine nitrogen increases the likelihood of nucleophilic attack. In very strongly basic conditions, substitution of halogenated pyridines occurs via a pyridyne intermediate such as (33) and a mixture of substituted products is obtained as in reaction (6.33).

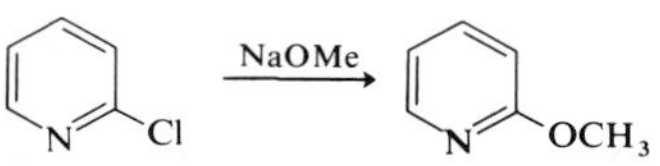

(6.31)

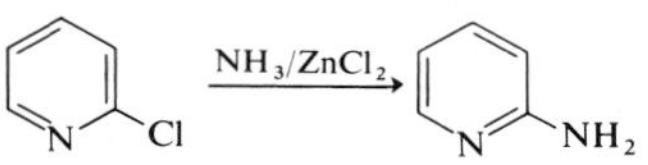

(6.32)

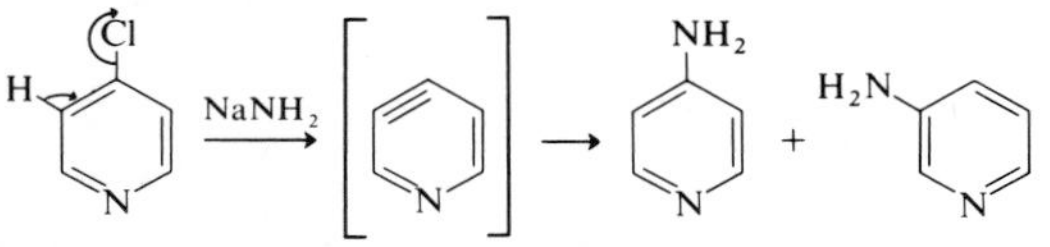

(6.33)

(33)

Benzenediazonium salts are prone to nucleophilic substitution and are thus synthetically useful. It is not surprising that 3-aminopyridine can be diazotised and made to undergo reactions such as the Sandmeyer reaction. 2- and 4-Aminopyridines, however, give diazonium salts which are too reactive to be useful.

The effect of additional heteroatoms is to make the carbon atoms of the ring even more electron-deficient than they are in pyridine and diazines are readily reduced and undergo nucleophilic substitution reactions as outlined in reactions (6.34) to (6.37) below. Interestingly, it has been shown recently by radioactive tracer techniques that substitution of a 6-bromo-pyrimidine is not straightforward but follows a pathway of the type outlined in (6.38) below.

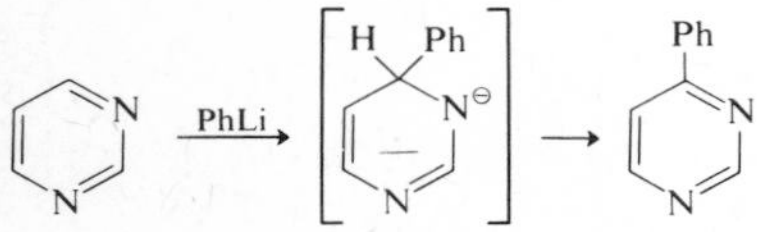

(6.34)

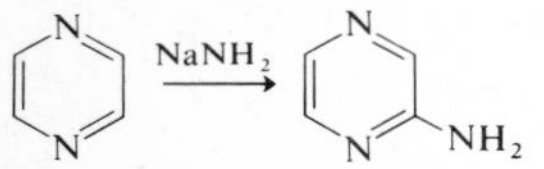

(6.35)

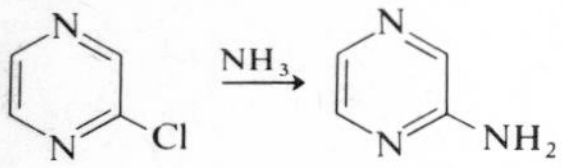

(6.36)

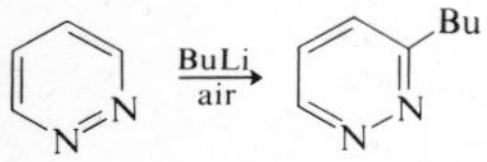

(6.37)

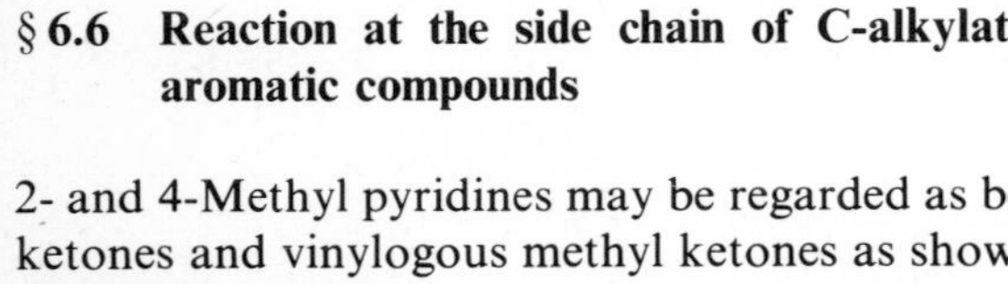

(6.38)

§ 6.6   Reaction at the side chain of C-alkylated six-membered hetero-aromatic compounds

2- and 4-Methyl pyridines may be regarded as being analogous to methyl ketones and vinylogous methyl ketones as shown in Fig. 6.1. The methyl groups will therefore be acidic, and carbanions may be formed in base so that condensation reactions may be achieved as in reactions (6.39) and (6.40). Activation of 3-alkyl substituents can only occur by inductive effects but, under forcing conditions, condensations such as (6.41) may be achieved.

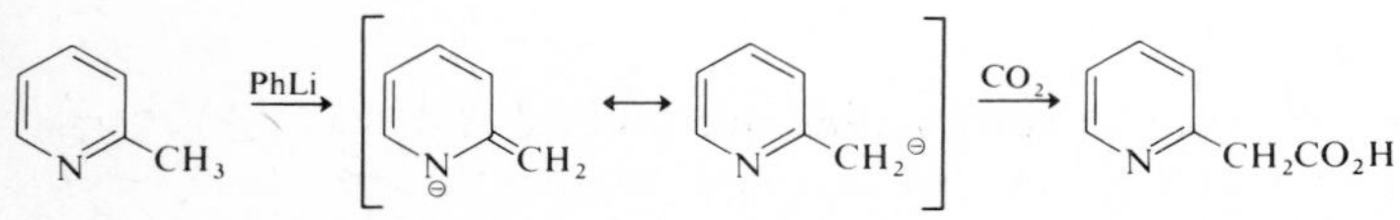

(6.39)

Figure 6.1

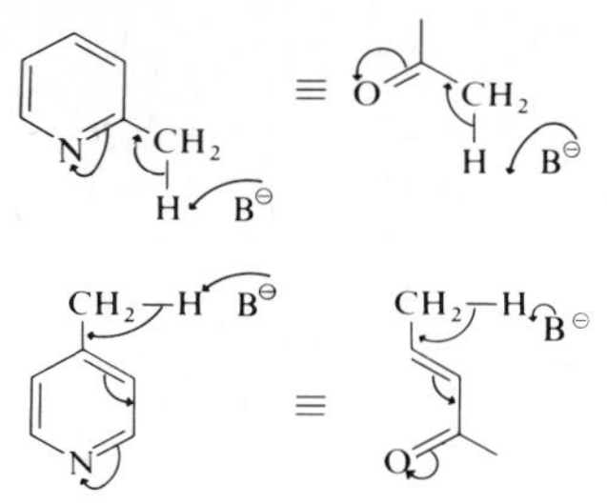

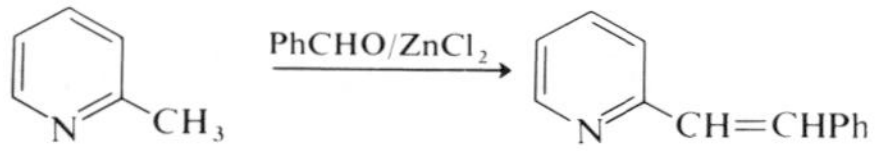

(6.40)

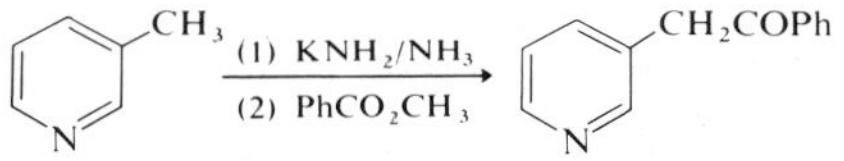

(6.41)

The reactivity of 2- or 4-alkylpyridines will be enhanced if the hetero-atom is made more electronegative by conversion to the corresponding pyridinium salt or N-oxide and condensations will occur under milder conditions as in reaction (6.42). It is probable that in reaction (6.40) the zinc chloride catalyst complexes at nitrogen and so has an effect similar to quaternisation.

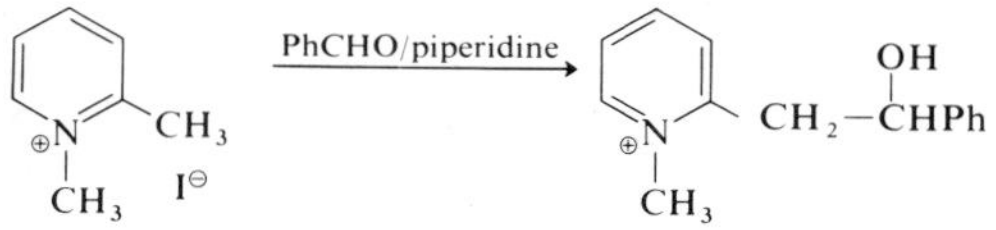

(6.42)

The analogy between alkyl pyridines and ketones is also evident in the Michael reactions such as (6.43) which are undergone by 2-vinyl-pyridine.

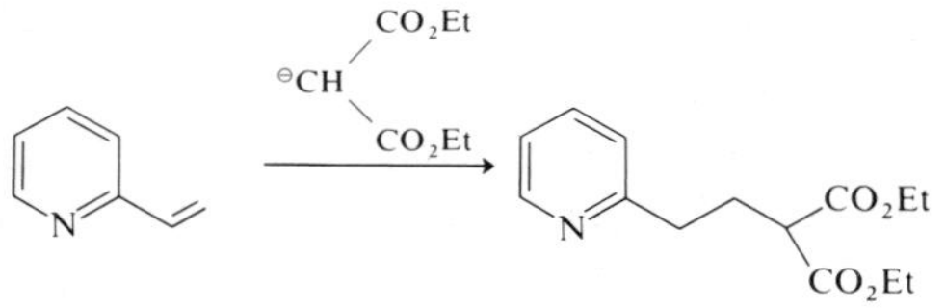

(6.43)

Alkyl diazines are activated in much the same way as are alkyl pyridines as witnessed by reactions (6.44) and (6.45). Alkylpyrilium salts may undergo deuterium exchange at 2- and 4-methyl substituents as in (6.46).

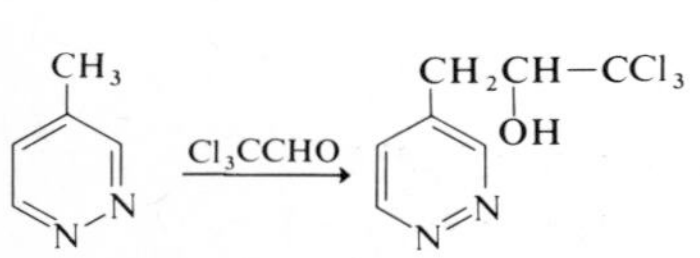

(6.44)

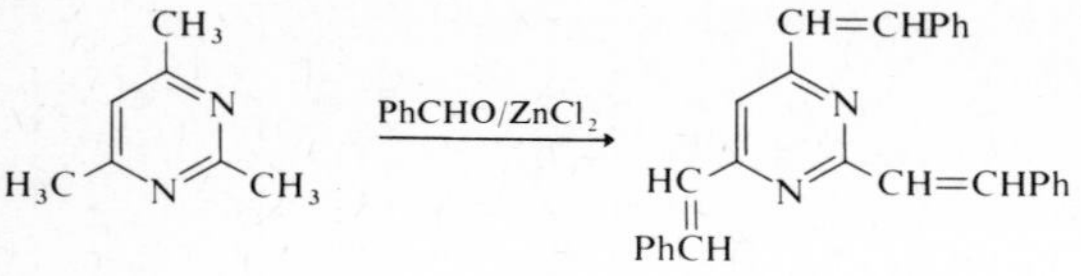

(6.45)

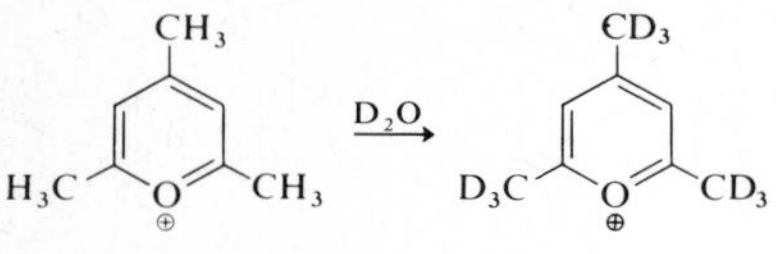

(6.46)

§ 6.7 Hydroxy and amino substituted six-membered heteroaromatic and related compounds

In Chapter 4 we have seen how 2- and 4-hydroxypyridines exist as the amide, or pyridone tautomers and yet because of amide resonance they may be considered to be aromatic with contributions from forms (34b) or (35b). We have also seen how other hydroxylated six-membered heteroaromatic compounds exist as amides or vinylogous amides.

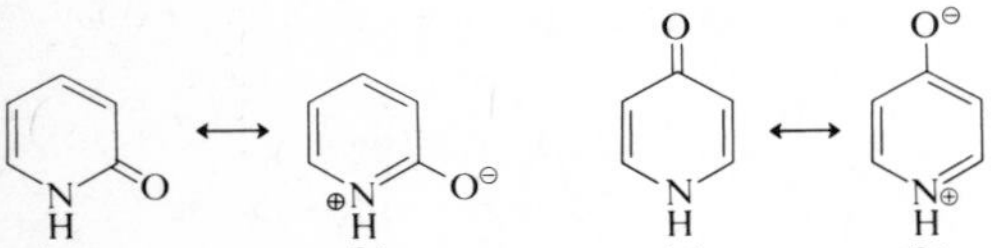

2-Pyridone does show some phenolic properties while the 4-isomer is less obviously phenolic. Both 2- and 4-pyridones, like most amides, can be N-alkylated but they also exhibit aromatic properties in being prone to substitution by electrophilic reagents. Reactions such as halogenation, nitration and sulphonation are undergone by these compounds. NMR work has suggested that 2-pyridone has 35 per cent aromatic character and it is perhaps significant that N-methyl-2-pyridone does not undergo Diels–Alder reactions while the less aromatic 2-pyrone does. Dipole moment studies indicate contributions from forms (34b) and (35b).

2- and 4-Pyrones might be expected to have some aromatic character due to the resonance forms (36b) and (37b). Since oxygen is more electronegative than nitrogen, its lone-pair is more localised and so we would not expect 2- and 4-pyrones to be as aromatic as 2- and 4-pyridones.

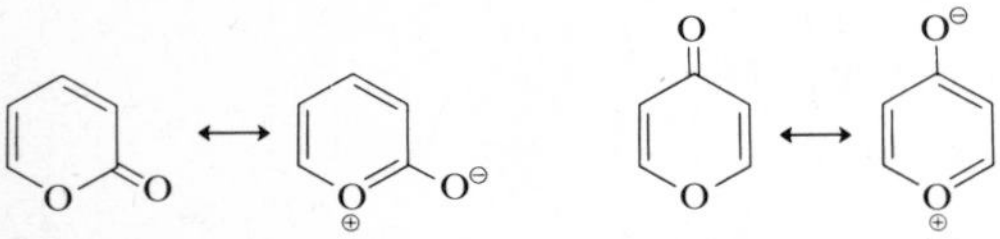

Some data, such as dipole moment studies, have suggested a slight degree of aromatic character in 2- and 4-pyrones and the halogenation reaction (6.47) seems to be an example of electrophilic aromatic substitution. Raman spectroscopy suggests that salts of 4-pyrones are phenolic and 4-pyrones will alkylate on oxygen as in reaction (6.48).

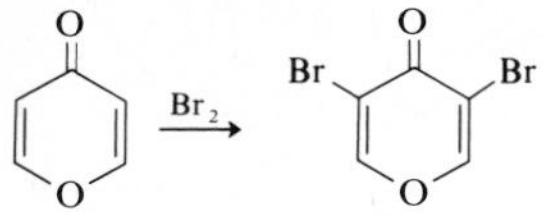

(6.47)

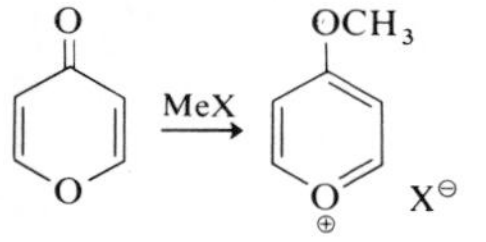

(6.48)

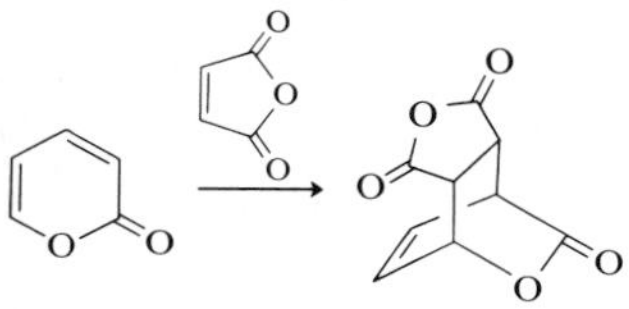

(6.49)

Recent magnetic susceptibility studies suggest that 2- and 4-pyrones have little or no aromatic character and the ease with which 2-pyrone undergoes the Diels–Alder reaction (6.49) is indicative of this.

Thiapyrones seem to have more aromatic character than pyrones, and perhaps because of this, oxidation of thiapyrone (38) to the sulphone (39) is much more difficult than oxidation of the reduced compound (40). It is of interest that, while the sulphone (39) exhibits ketonic properties, the thiapyrone (38) does not.

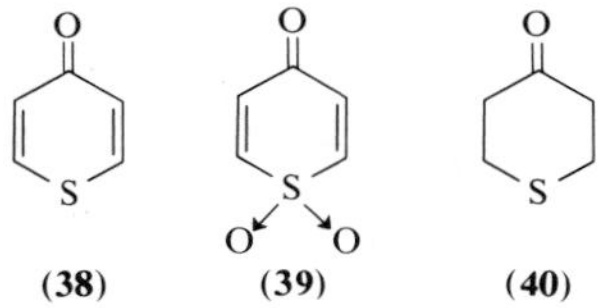

(38)   (39)   (40)

As we have seen in Chapter 4, 2- and 4-aminopyridines and analogous compounds exist in the amino form although dipole moment studies suggest that there is some contribution from the canonical forms (21b) and (41b).

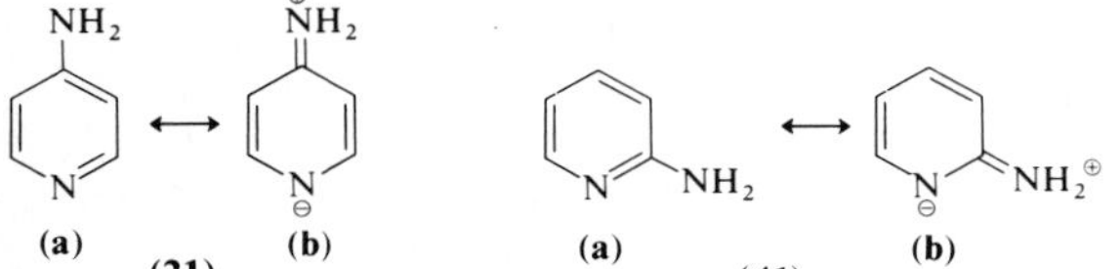

(a)   (b)   (a)   (b)
     (21)   (41)

§ 6.8   Other six-membered heteroaromatic compounds

Recently there has been some interest in heterocyclic analogues of the compound (42) which has special stability as the dipolar form (42b) due to the aromatic character of both cyclopentadienyl anion and tropylium cation components. The compound (43) is isoelectronic with (42) and physical studies indicate that when X = O, the non-polar form (43a) predominates with some contribution from (43b).

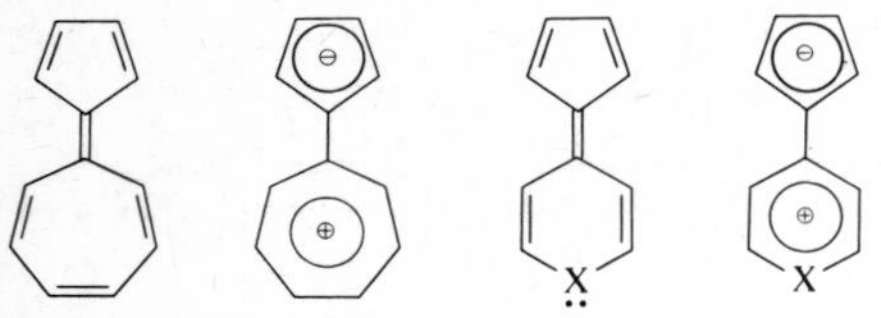

(42a)   (42b)   (43a)   (43b)

§ 6.9   Synthesis of six-membered heteroaromatic compounds

The *Hantzsch Synthesis* of symmetrically substituted pyridines is very widely used. It involves reaction of a β-ketoester with an aldehyde in the presence of ammonia as in (6.50). The product, a dihydropyridine, can be oxidised to the pyridine. 1,5-Diketones will react with either ammonia to yield dihydropyridines or with hydroxylamine to yield the pyridines directly as in (6.51) and (6.52).

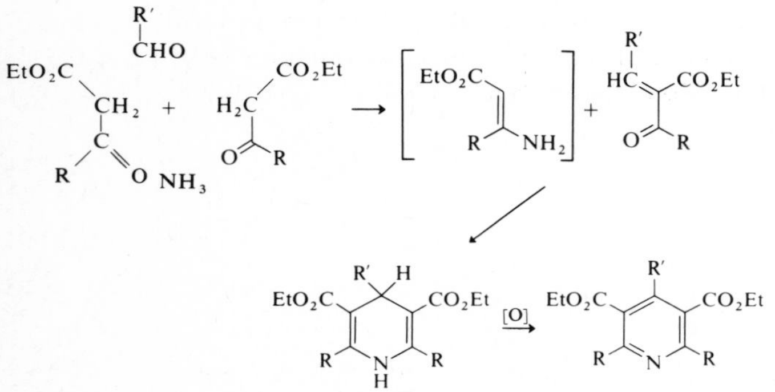

(6.50)

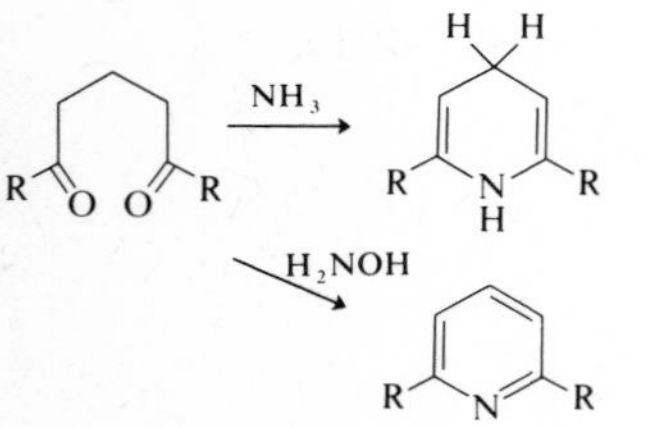

(6.51)

(6.52)

Guareschi synthesis involving reaction of cyanoacetamide with a diketone yields pyridones which can be converted to pyridines as in (6.53). The self-condensation of acetoacetic ester (6.54) may be regarded as a variant of this reaction and the 2-pyrone dehydroacetic acid (44) can rearrange to a 4-pyrone in sulphuric acid.

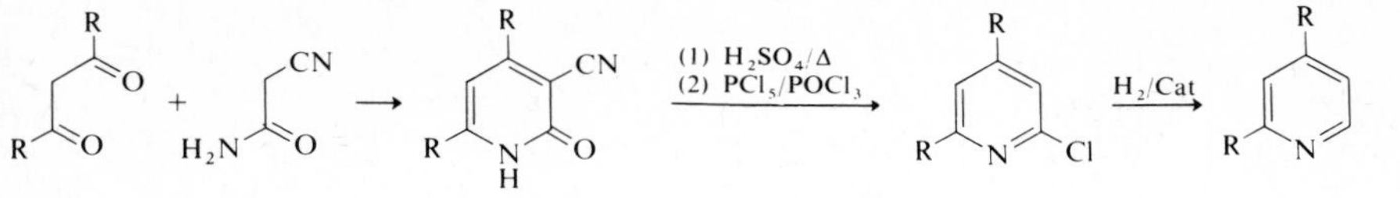

(6.53)

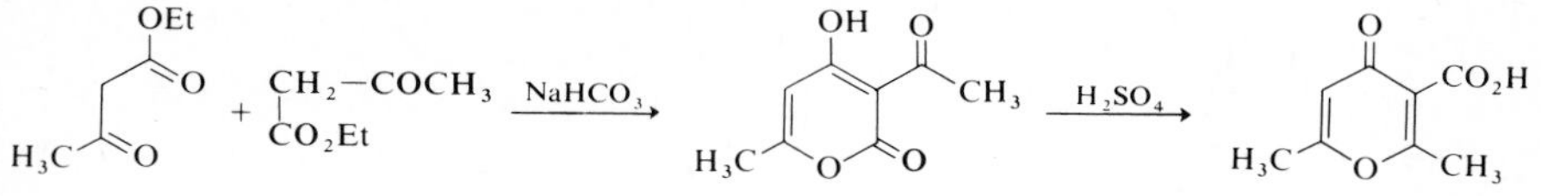

(6.54)

(44)

Just as pyridine can be made from the reaction of 1,5-diketones with ammonia, pyridazines may be made from the reaction of 1,4-diketones and hydrazine as in (6.55) and pyrimidines can be made from 1,3-diketones and urea as in (6.56). Pyrazines are made by self-condensation of α-amino ketones which are prepared *in situ* as shown in (6.57) or by condensation of 1,2-diketones with 1,2-diamines as in (6.58).

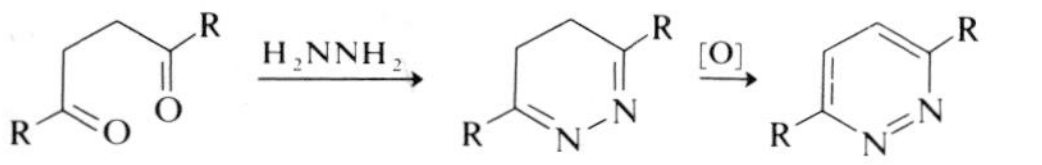

(6.55)

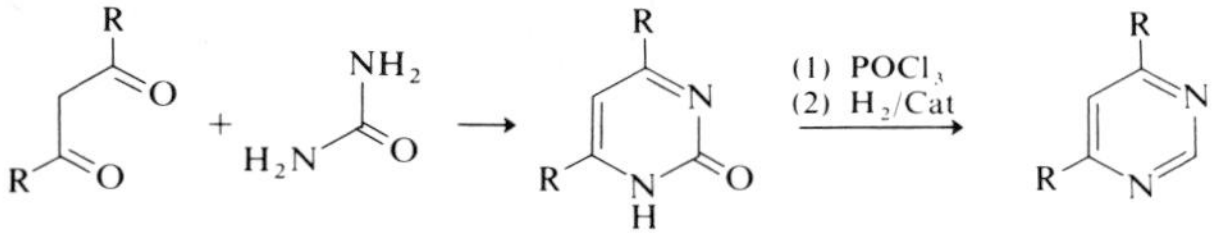

(6.56)

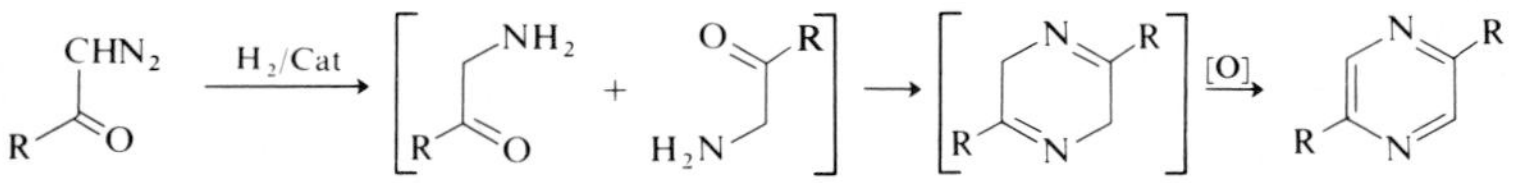

(6.57)

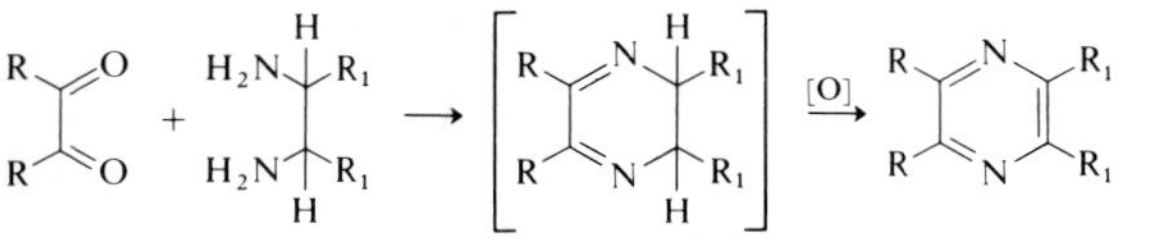

(6.58)

A general method for making compounds in this series which have heteroatoms other than nitrogen has been developed by Ashe and involves organotin intermediates:

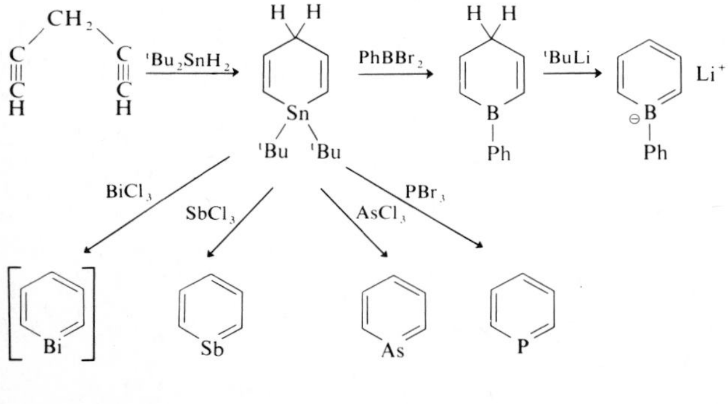

Key points

(1) Six-membered heteroaromatic compounds are structurally similar to benzene, and the more different the structure is from that of benzene, the less aromatic are the properties of the compound.

(2) The heteroatoms, by withdrawing electron density from the ring carbon atoms, make typically benzenoid reactions such as electrophilic aromatic substitution difficult, although such synthetically useful reactions may be undergone by N-oxides.

(3) Nucleophilic substitution reactions which are not common in benzene chemistry may be achieved with six-membered heteroaromatic compounds. Quaternary salts are especially prone to undergo such reactions.

(4) Halogen groups α- and γ- to nitrogen atoms are readily substituted by nucleophilic reagents. Such nucleophilic substitution reactions are not common with phenyl halides.

(5) Alkyl substituents which are α- and γ- to the heteroatoms in such compounds have acidic hydrogens and so can participate in reactions with electrophilic reagents. Quaternisation enhances the acidity of α- and γ-alkyl hydrogens.

Further reading

General
R. C. Elderfield, ed. *Heterocyclic Compounds*, vol. 1, Wiley, New York, 1950; *ibid.*, vol. 6, 1957.

Pyridine
E. Klingsberg, ed. *Pyridine and its Derivatives*, parts 1–4, Interscience, New York, 1960–1964.
R. A. Abramovitch and J. G. Saha, *Advances in Heterocyclic Chemistry*, 1966, **6**, 229, Academic Press, New York.

Pyrilium Salts
A. T. Balaban, W. Schroth and G. Fischer, *Advances in Heterocyclic Chemistry*, 1969, **10**, 241, Academic Press, New York.

Diazines
M. Tisler and B. Stanovnik, *Advances in Heterocyclic Chemistry*, 1968, **9**, 211, Academic Press, New York (pyridazines).
D. J. Brown, *The Pyrimidines*, Interscience, New York, 1962.
D. J. Brown, *The Pyrimidines Supplement I*, Wiley-Interscience, New York, 1970.

Nucleophilic Heteroaromatic Substitution
G. Illuminati, *Advances in Heterocyclic Chemistry*, 1964, **3**, 285, Academic Press, New York.

Chapter 7

Polycyclic aromatic heterocyclic compounds

When the five- and six-membered heteroaromatic compounds which have been discussed in Chapters 5 and 6 are fused together in polycyclic systems, the resultant heterocyclic compounds may have quite different properties from those of the "parent" compounds from which they derive. There is a large number of compounds of this type and many are important in biological systems.

§ 7.1 Bicyclic compounds derived from six-membered heteroaromatic compounds

There are three ways in which pyridine may be fused with benzene; the resultant heterocyclic compounds are quinoline (**1**), isoquinoline (**2**) and quinolizinium salts (**3**).

For quinoline and isoquinoline, resonance forms (**1a–1e**) and (**2a–2e**) respectively may be written. The charged forms (**1d**) and (**1e**), and (**2d**) and (**2e**) are given credence by dipole moments of 2·1 D for quinoline and 2·6 D for isoquinoline.

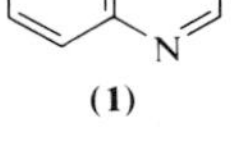

(1) 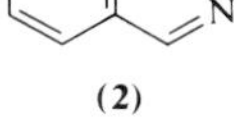   (2) 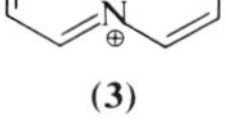   (3)

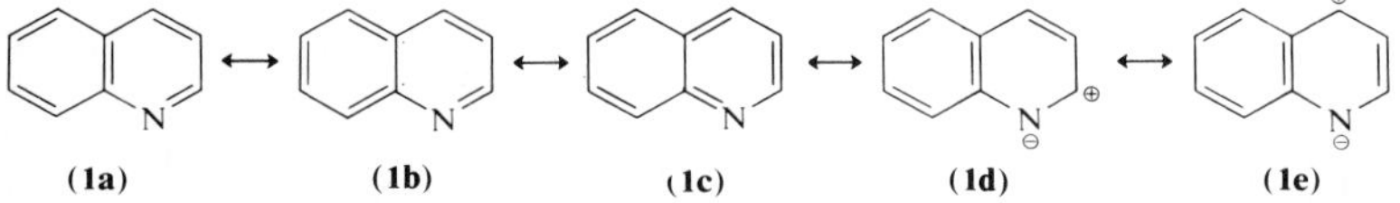

(1a)   (1b)   (1c)   (1d)   (1e)

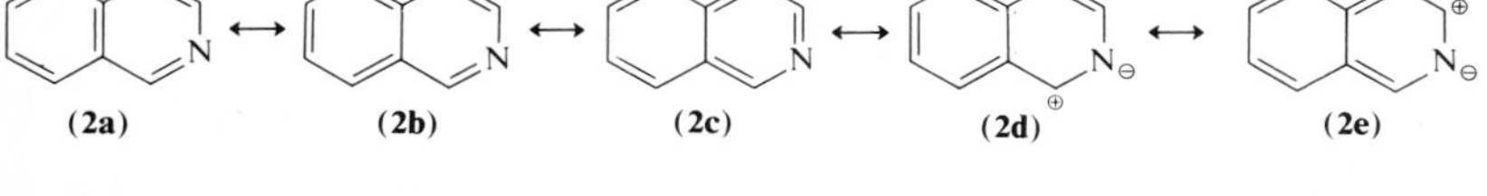

(2a)   (2b)   (2c)   (2d)   (2e)

In both quinoline and isoquinoline the "pyridine" ring is π-deficient, and, since oxidation is dependent on electron availability, permanganate oxidation will destroy the "benzene" ring and leave the "pyridine" ring intact as in reactions (7.1) and (7.2).

$$\text{quinoline} \xrightarrow{\text{KMnO}_4} \quad \text{pyridine-3,4-dicarboxylic acid (HO}_2\text{C, HO}_2\text{C)} \qquad (7.1)$$

$$\text{isoquinoline} \xrightarrow{\text{KMnO}_4} \quad \text{(HO}_2\text{C, HO}_2\text{C) pyridine derivative} \qquad (7.2)$$

§ 7.1.1   Reactions at nitrogen

Quinoline and isoquinoline have $pK_a = 4.94$ and 5.4 respectively and so have similar base strengths to pyridine. Quaternary salts and N-oxides can be formed in much the same way as the corresponding pyridine derivatives are formed (§ 6.2.2).

§ 7.1.2   Electrophilic aromatic substitution

Since the pyridine ring is π-deficient and can be made even more electron-deficient by protonation or quaternisation, electrophilic reagents will show a preference for attack on the benzene ring of quinoline and iso-quinoline, especially if acidic conditions are used. Quinoline is nitrated at C_5 and C_8 as in (7.3) below and sulphonated at C_8 as in (7.4) below. Both of these reactions occur in strongly acidic media, although at low acid strengths, substitution in the pyridine ring at C_3 may be observed as in reaction (7.5). Isoquinoline is nitrated and sulphonated at C_5 (cf. (7.6) below).

$$\text{quinoline} \xrightarrow{\text{HNO}_3/\text{H}_2\text{SO}_4} \quad \text{5-nitroquinoline (NO}_2\text{)} + \text{8-nitroquinoline (NO}_2\text{)} \qquad (7.3)$$

$$\text{quinoline} \xrightarrow{\text{H}_2\text{SO}_4} \quad \text{quinoline-8-sulphonic acid (SO}_3\text{H)} \qquad (7.4)$$

$$\text{quinoline} \xrightarrow{\text{HNO}_3/\text{Ac}_2\text{O}} \quad \text{3-nitroquinoline (NO}_2\text{)} \qquad (7.5)$$

$$\text{isoquinoline} \xrightarrow{\text{H}_2\text{SO}_4} \quad \text{isoquinoline-5-sulphonic acid (SO}_3\text{H)} \qquad (7.6)$$

Just as strongly acidic conditions can deactivate a pyridine ring towards electrophilic reagents, so formation of N-oxides will activate a pyridine ring towards electrophilic reagents. Quinoline-N-oxide may therefore be substituted by electrophilic reagents in the "pyridine" ring as in (7.8) below. The precise reaction conditions are, however, important in determining the nature of the product, and in reactions (7.7) and (7.9) below, conjugate acid formation may be responsible for directing nitration to the benzene ring.

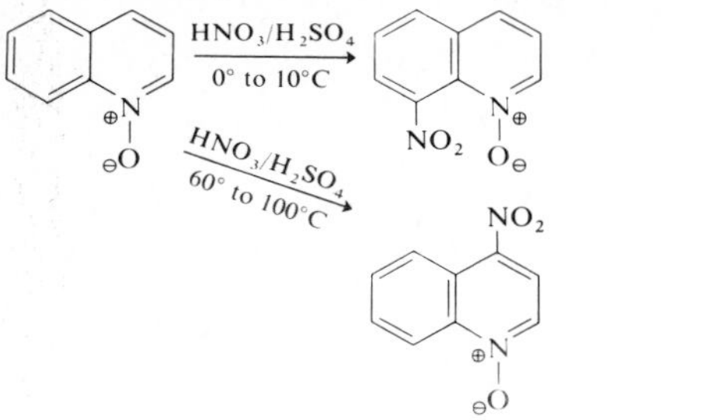

(7.7)

(7.8)

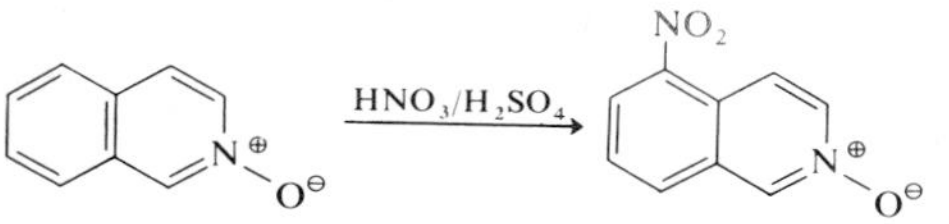

(7.9)

§ 7.1.3 Nucleophilic aromatic substitution

This will occur in the "π-deficient" pyridine ring and, in quinoline, C_2 is the preferred site of attack, while, in isoquinoline, attack occurs at C_1. This substitution pattern reflects the importance in quinoline of the canonical form (1d) which has styrene conjugation (which canonical form (1e) lacks). The canonical form (2d) for isoquinoline retains the benzene ring as an aromatic species ((2e) does not) and so makes for C_1 substitution by nucleophilic reagents. When the preferred positions are occupied by substituents then C_4 in quinoline and C_3 in isoquinoline may be attacked. The Tschitschibabin reactions (7.10) and (7.11) are examples of normal attack as are reaction (7.12) and the addition reaction (7.13).

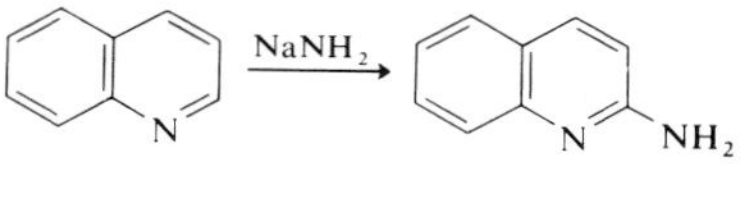

(7.10)

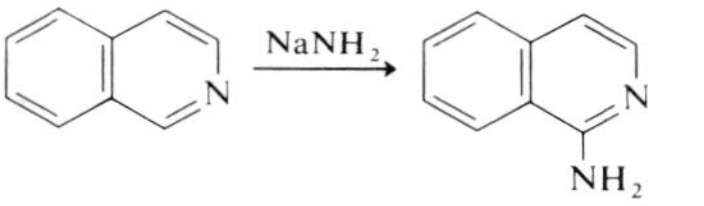

(7.11)

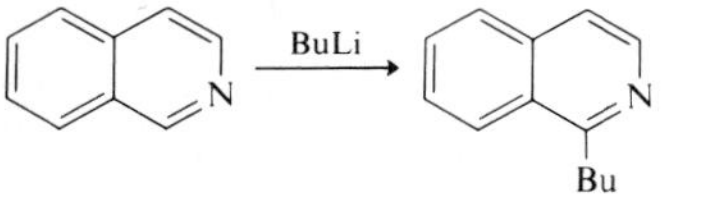

(7.12)

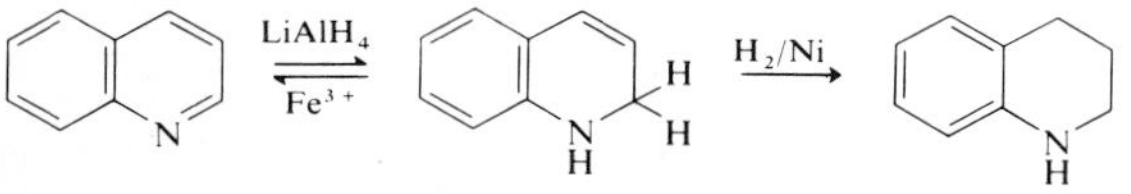

(7.13)

Quaternisation reinforces the ability of pyridines to react with nucleophilic reagents (cf. § 6.5) and so quinoline and isoquinoline can react very readily with nucleophilic reagents via quaternary salts. Examples of such reactions are (7.14), (7.15) and (7.16), and the Reissert reaction (7.17).

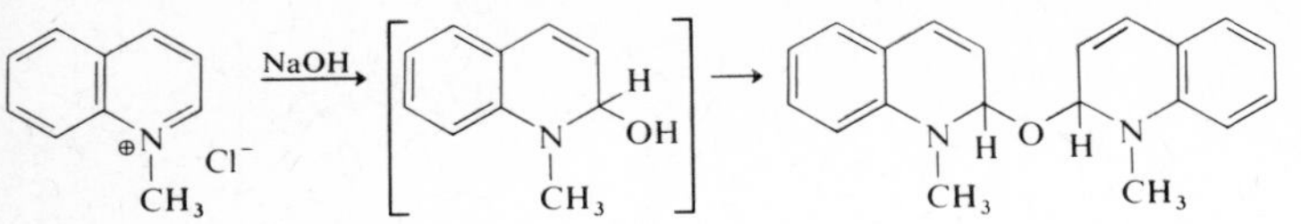

(7.14)

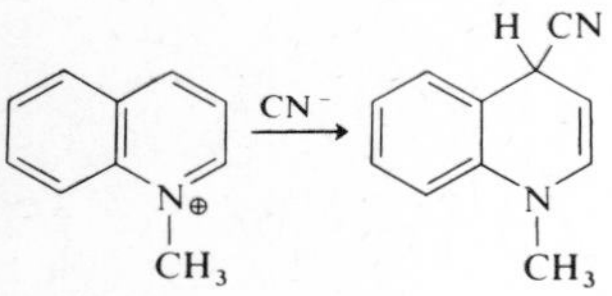

(7.15)

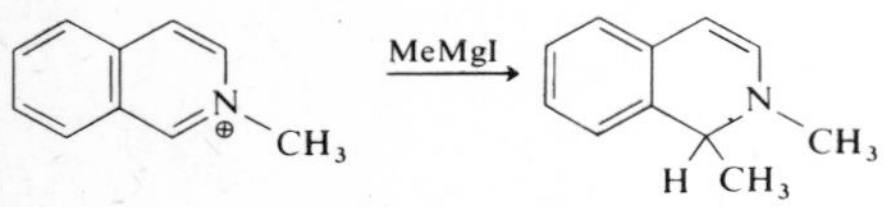

(7.16)

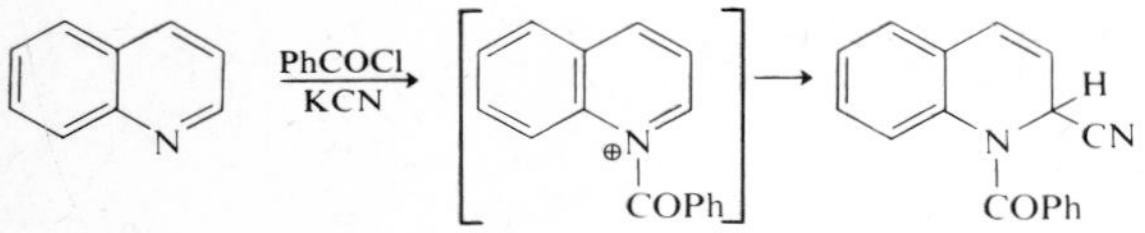

(7.17)

A general method of preparing alkyl and alkenyl substituted heterocyclic compounds is to react the chloro-heterocyclic compound with a Wittig reagent as in (7.18) below. The intermediate ylide may then be hydrolysed with base to give the alkyl derivative or it may be reacted with a carbonyl compound to yield an alkenyl derivative as shown.

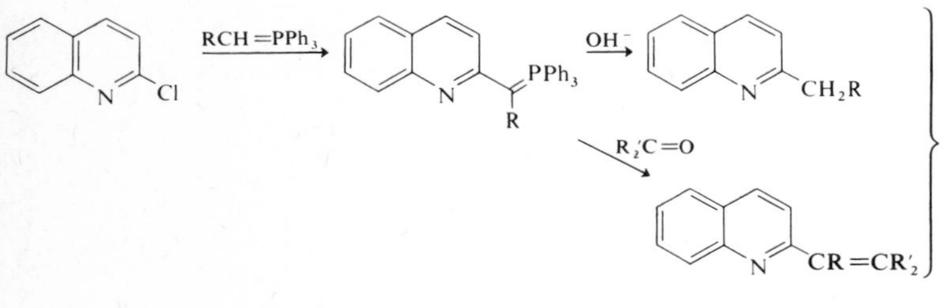

(7.18)

2- and 4-Chloroquinolines and 1- and 3-chloroisoquinolines, like 2- and 4-chloropyridines, are readily attacked by nucleophilic reagents with replacement of the halogen atom as in (7.19) and (7.20) below. The chloro compounds, which are readily obtained from the quinolones and isoquinolones by reaction with phosphorus halides, are extremely useful intermediates in synthesis of substituted quinolines and isoquinolines. The order of reactivity is 2-chloroquinoline > 4-chloroquinoline; and 1-chloroisoquinoline > 3-chloroisoquinoline (cf. (7.21) below).

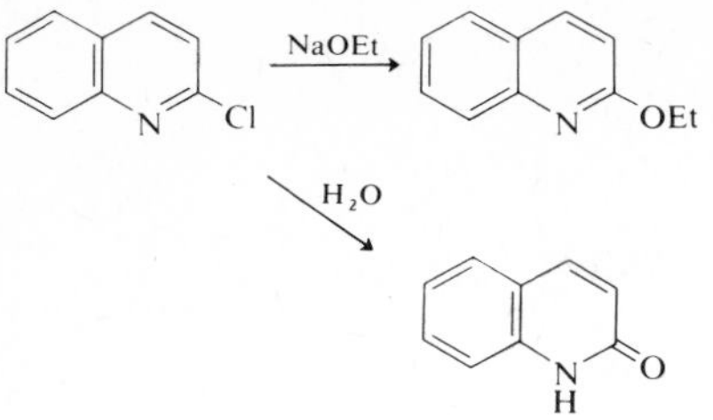

(7.19)

(7.20)

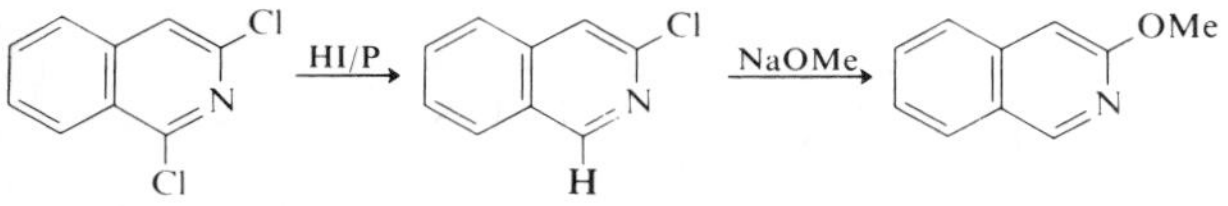

(7.21)

§ 7.1.4   Hydroxy- and amino-quinolines and isoquinolines

As expected from Chapter 4, 2- and 4-hydroxyquinolines (**4**) and (**5**) and 1- and 3-hydroxyisoquinolines (**6**) and (**7**) occur as the amide tautomers. The 3-isoquinolone (**7**) can only be written as the quinonoid structure which might imply loss of the benzene aromaticity. There is, however, no evidence that the aromatic ring here in any way differs from that in any of the other compounds.

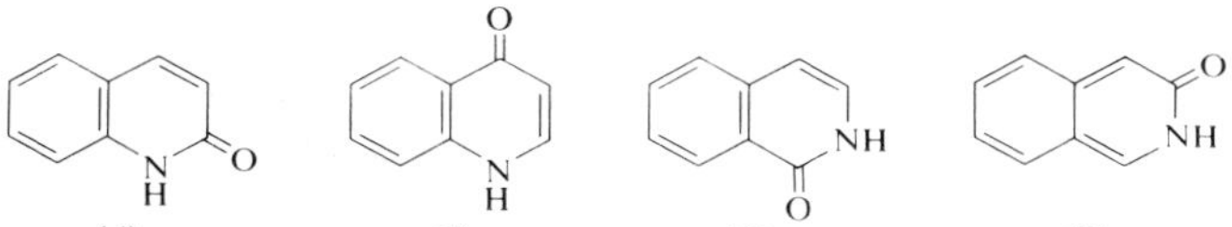

All other hydroxyquinolines and isoquinolines are in the phenol form, and all the amino compounds are in the amino form, as might be expected in view of the discussion in Chapter 4 (§ 4.4).

§ 7.1.5   Reactivity of alkyl-substituted compounds

Like α- and γ-methyl pyridines (§ 6.6), 2- and 4-alkyl quinolines and 1- and 3-alkyl isoquinolines have acidic hydrogens and can react with aldehydes and ketones in basic conditions. Activation is greatest in 2-methyl-quinoline and in 1-methylisoquinoline; (7.22) and (7.23) are typical reactions of these compounds.

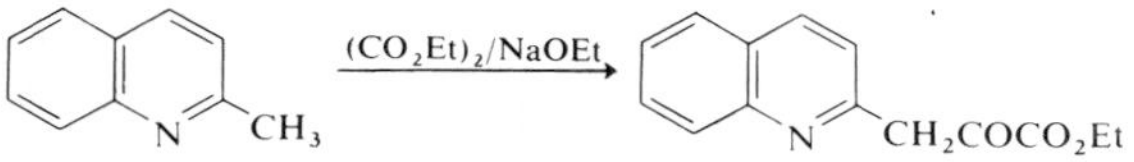

(7.22)

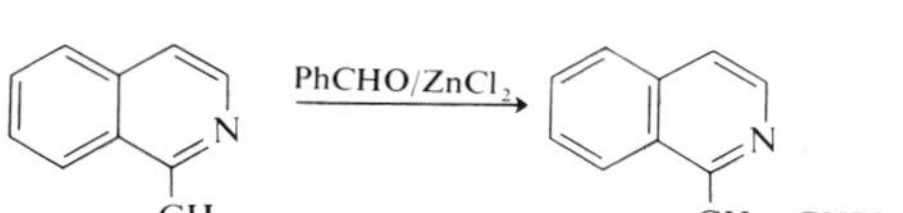

(7.23)

§ 7.1.6   Quinolizinium salts

These (general structure **3**) are similar to pyridinium salts in their reactions. Nucleophilic reagents can react with ring opening as in (7.24) and (7.25) and methyl substituents which are α- and γ- to the quaternary nitrogen are activated towards condensation reactions such as reaction (7.26).

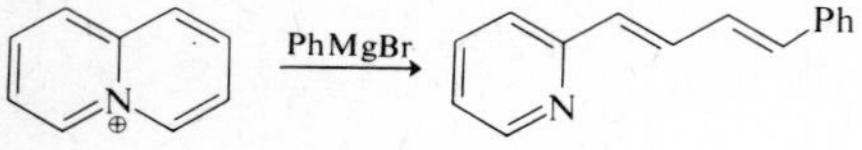

(7.24)

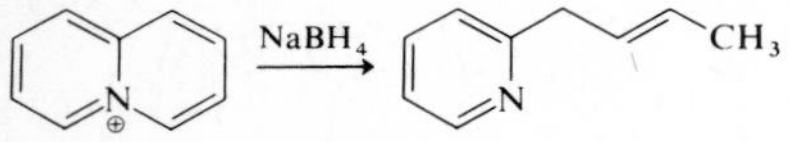

(7.25)

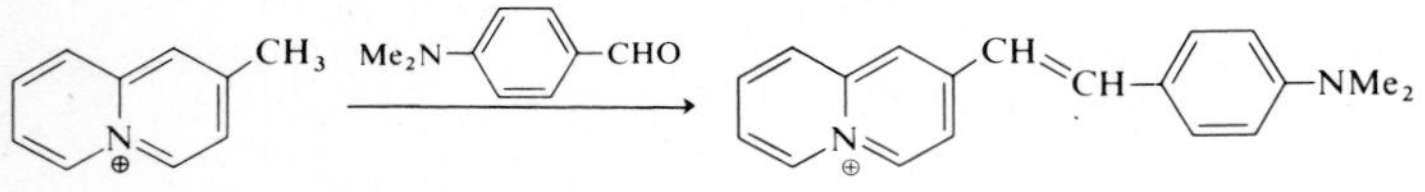

(7.26)

§ 7.1.7   Bicyclic compounds with more than one heteroatom in one ring

As we have seen in Chapter 6, additional heteroatoms serve further to deactivate a pyridine ring towards electrophilic reagents. Attack of such reagents will, therefore, be even more specific to the benzene ring of those bicyclic heterocyclic compounds with more than one nitrogen in one of the rings. The heterocyclic ring will be activated towards nucleophilic reagents and alkyl groups α- and γ- to nitrogen atoms will be activated for condensation with aldehydes and ketones. Hydroxyl groups α- or γ- to nitrogen atoms again occur as the keto (or amide) tautomer.

Examples of compounds with two nitrogen atoms in the same ring are cinnolines (**8**), phthalazines (**9**), quinazolines (**10**) and quinoxalines (**11**). These compounds are all more weakly basic than quinoline due to the electron-withdrawing effect of the second heteroatom, and will form N-oxides and quaternary salts. Permanganate oxidation occurs in much the same way as in reactions (7.1) and (7.2) and some typical reactions are shown in reactions (7.27) to (7.30).

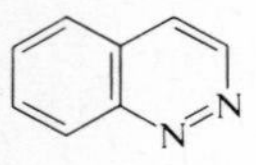

(**8**) 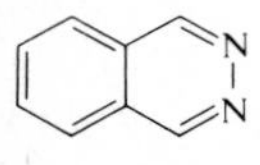   (**9**) 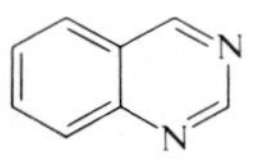   (**10**) 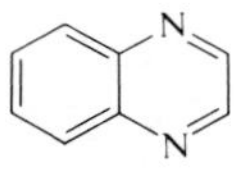   (**11**)

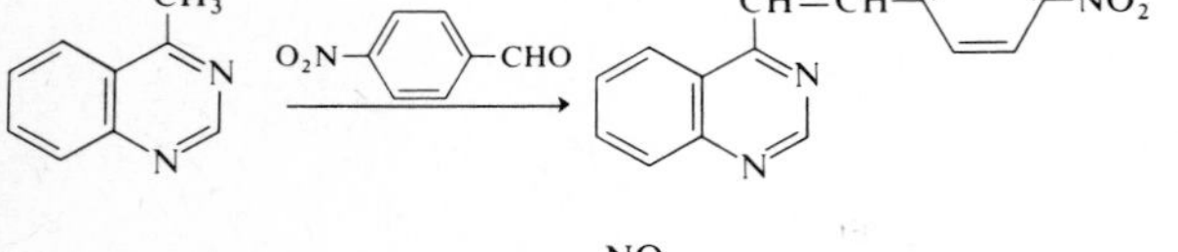

(7.27)

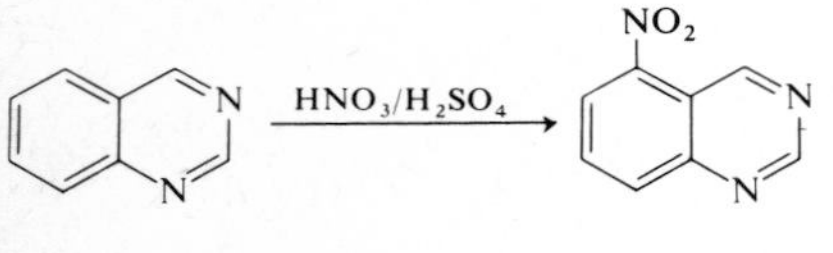

(7.28)

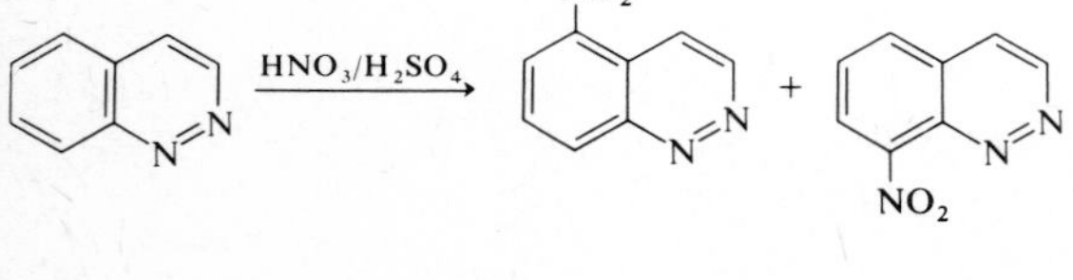

(7.29)

(7.30)

(7.31)

(12)   **(13)**

The weakening of the aromaticity in quinoxalines is shown by the occurrence of the Diels–Alder reaction (7.31). This would imply that there is a contribution from the quinomethine form **(13)** in 2,3-dimethyl-quinoxaline **(12)**.

§ 7.1.8 *Bicyclic six-membered aromatic compounds with heteroatoms in both rings*

Examples of all six naphthyridines, **(14)** to **(19)**, are known and the reactivity can be predicted by extrapolation from the chemistry of quinolines and other compounds in this chapter. The compounds are resistant to attack by electrophilic reagents. Alkyl groups α- and γ- to the heteroatom are activated for condensation with aldehydes, and chloro groups similarly situated are readily replaced by nucleophilic reagents such as NaOR, NH_3, RNH_2, etc. Reactions at nitrogen such as quaternisation by alkyl halides, and oxidation to N-oxides are known.

(14)   **(15)**   **(16)**   **(17)**   **(18)**   **(19)**

Pteridines with the basic structure **(20)** were first found in the pigments of butterfly wings, and are an extremely important class of polyaza-naphthalenes, Since there are four nitrogen atoms in the molecule, pteridines are extremely π-deficient and so electrophilic aromatic sub-stitution reactions are unknown. The localisation of electron density at nitrogen causes a considerable reduction in the aromaticity of the pteridine system and so hydrolytic cleavage of unsubstituted pteridines in both acidic and basic conditions can occur as in reaction (7.32). The pyrimidine ring is cleaved in preference to the pyrazine ring.

(7.32)

(20)

Strongly electron-donating substituents can compensate for the electron deficiency of pteridines and so the stability of naturally occurring pteridines is enhanced by substitution.

The electron deficiency of pteridines makes for easy nucleophilic substitution reactions, and, while chloro-substituted pteridines, like chloro derivatives of most π-deficient heteroaromatic compounds, can undergo ready substitution by a variety of nucleophilic reagents to yield many useful derivatives, even unsubstituted positions in pteridines may undergo nucleophilic substitution and addition reactions. Some of these reactions are exemplified by reactions (7.33) to (7.39) below. The easily reversible addition of the weak nucleophile water as in (7.37) yielding an "aliphatic" amino group, accounts for the fact that pteridine ($pK_a = 4.1$) is a stronger base than either pyrimidine ($pK_a = 1.3$) or pyrazine ($pK_a = 0.65$).

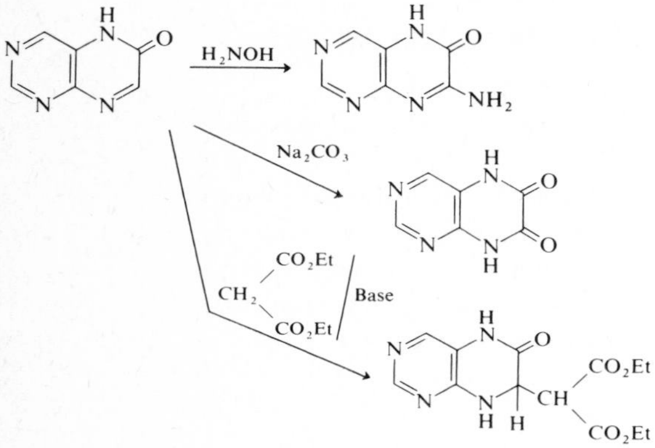

(7.33)

(7.34)

(7.35)

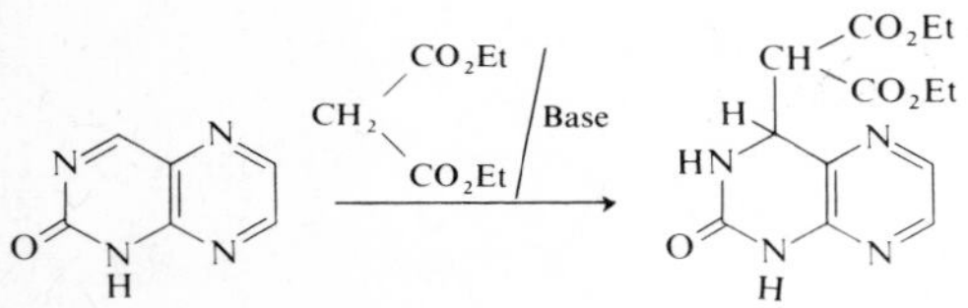

(7.36)

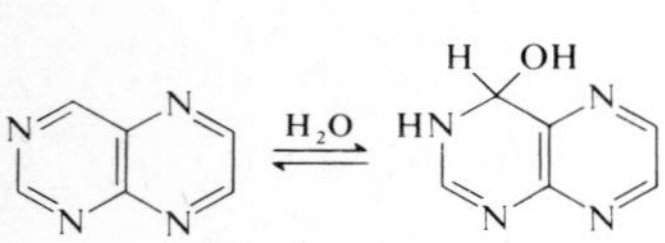

(7.37)

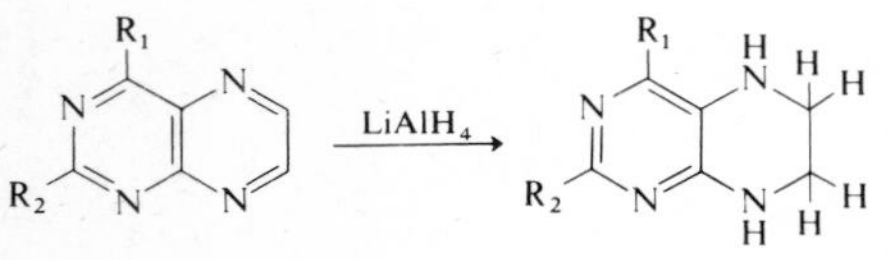

(7.38)

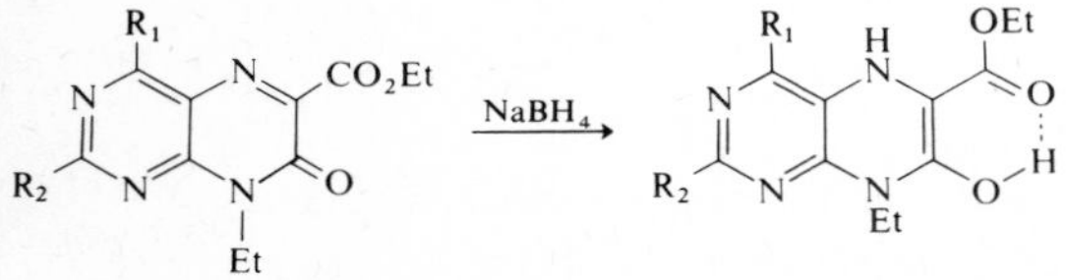

(7.39)

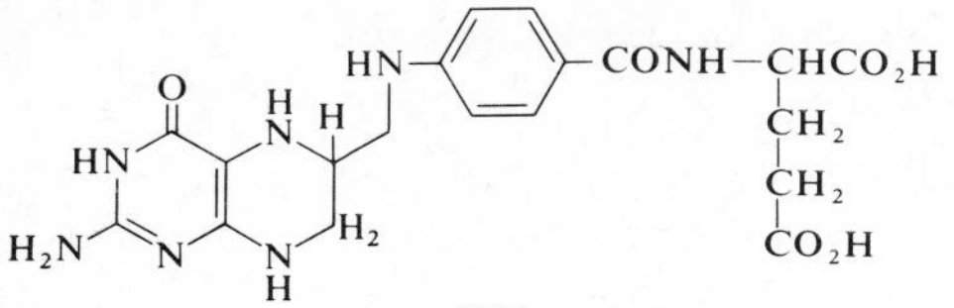

(21)

The reduced aromaticity of these systems is shown by the fact that autoxidation does not occur in reactions (7.35) and (7.36) to yield the "fully aromatic" systems. The nucleophilic reducing agents $LiAlH_4$ and $NaBH_4$ also give addition products as in (7.38) and (7.39) and reduced pteridines are extremely important in nature. The one-carbon transfer coenzyme tetrahydrofolic acid (21) is a reduced pteridine.

§ 7.1.9   Benzopyrilium salts, coumarins and chromones

Benzopyrilium salts (22) are similar to quaternary quinolinium salts and are susceptible to addition reactions in the heterocyclic ring at positions 2 and 4, as in reactions (7.40) and (7.41). As with quinolinium salts, methyl groups at positions 2 and 4 are activated towards condensation with aldehydes and ketones, as in reaction (7.42).

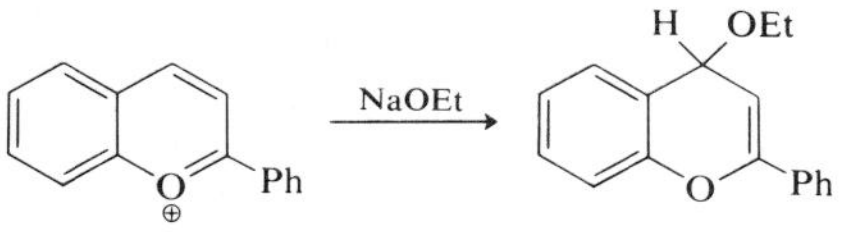

(22)   (7.40)

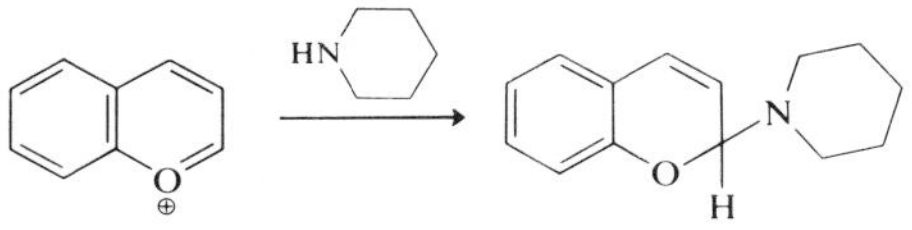

(7.41)

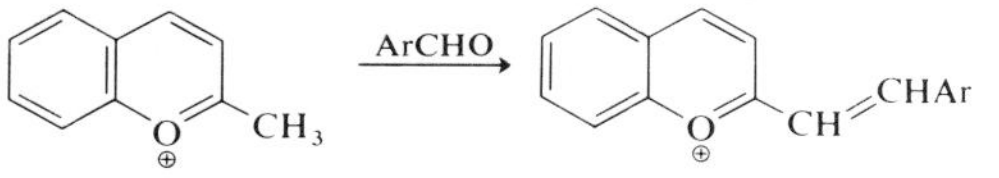

(7.42)

Coumarin (23) and chromone (24) may be regarded as hydroxybenzopyrilium salts. They are undoubtedly esters and vinylogous esters and coumarins undergo reactions typical of esters such as (7.43).

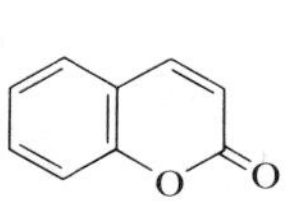

(23) 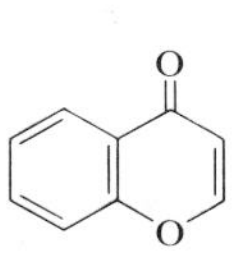   (24)

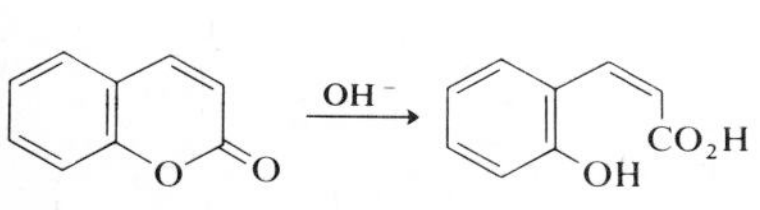

(7.43)

§ 7.2 Bicyclic compounds derived from five-membered heteroaromatic compounds

The common π-excessive heteroaromatic compounds can condense with benzene in two distinct ways yielding ((25) and (26), Z = NR, O and S). The resultant compounds, indole ((25), Z = NH), benzofuran ((25), Z = O), benzothiophene ((25), Z = S), isoindole ((26), Z = NH), isobenzofuran ((26), Z = O), and isobenzothiophene ((26), Z = S) have properties which are quite different from the parent compounds. Analogues involving other heteroatoms ((25), Z = PR, AsR, CrR and AlR) have been made.

The only resonance forms which can be written for the compounds in the series (25) which do not involve destruction of the benzenoid 6π-system are (25a) and (25b) and the high reactivity at position 3, compared to the high reactivity at C_2 of the parent series, is accounted for by this.

§ 7.2.1 Reactions of the heteroatom

Indole, benzofuran and benzothiophene are the most common compounds of this series and indole is by far the most studied of these. The base strength of indole ($pK_a = -3.5$) is very weak and nmr studies have shown that protonation occurs at C_3 to yield the ion (27). Grignard reagents will react with the acidic proton to give salts (cf. § 7.2.2). Benzothiophene can be oxidised to the sulphone as in (7.44) but in this case the sulphone is much more stable than thiophene sulphone (§ 5.3) and will only undergo the self Diels–Alder reaction (7.44) under forcing conditions.

(7.44)

§ 7.2.2 Electrophilic aromatic substitution reactions

These invariably occur at C_3 in indole and benzothiophene, as exemplified by the sulphonation reaction (7.45), the Vilsmeier reaction (7.46), the Mannich reaction (7.47), the Michael reaction (7.48) and the Friedel–Crafts and nitration reactions (7.49) and (7.50). Indole is autoxidised in base, as in reaction (7.51), by primary attack of the electrophilic oxygen at C_3 followed by nucleophilic attack of the peroxide anion on the resultant

Schiff's base to give a dioxetane intermediate. Fission of this dioxetane by the "forbidden" process shown can, in the presence of suitable compounds, result in chemiluminescence.

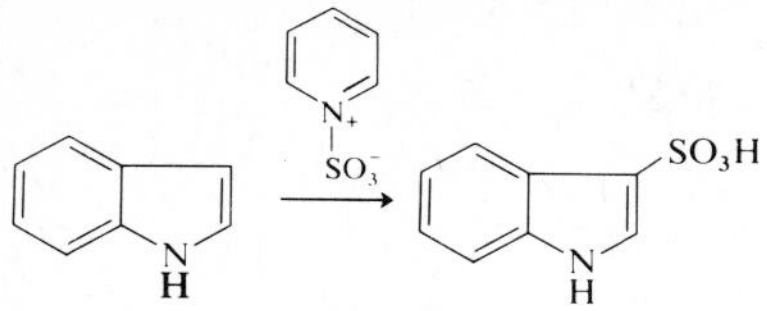

(7.45)

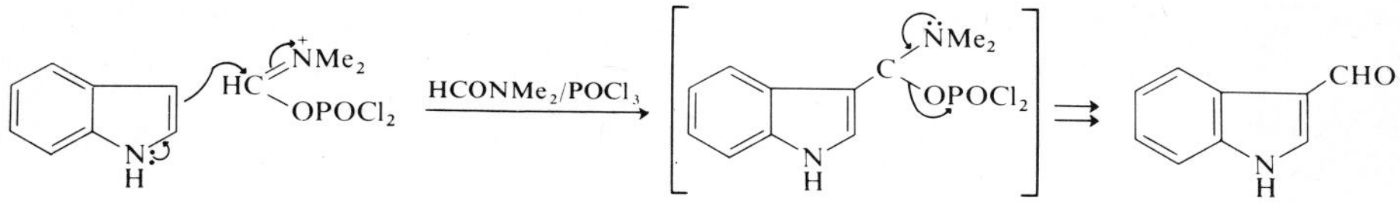

(7.46)

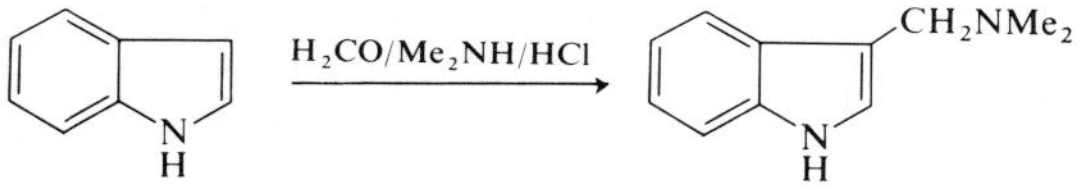

(7.47)

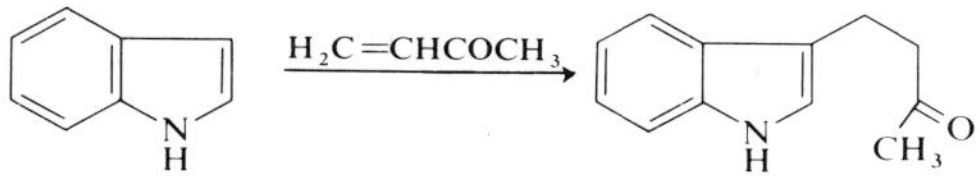

(7.48)

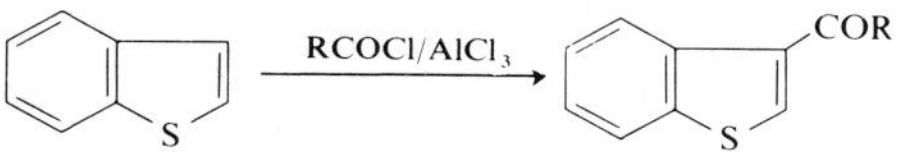

(7.49)

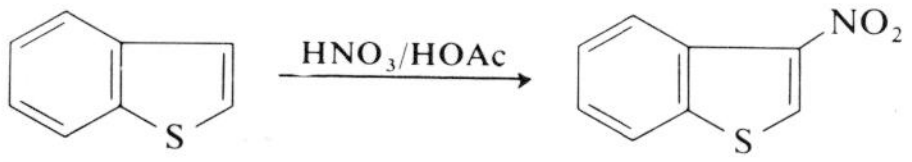

(7.50)

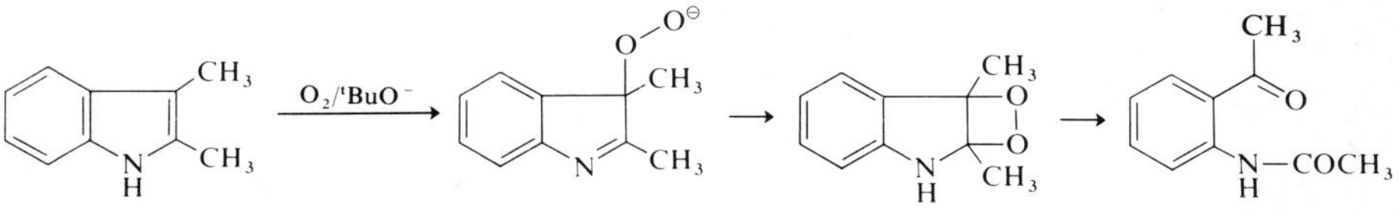

(7.51)

When C_3 of indole is occupied, electrophilic attack occurs at C_2 via the 3,3-disubstituted compound. Alkylation with excess methyl iodide eventually gives the tetramethyl derivative as in reaction (7.52). Benzofuran is atypical in its reactions and undergoes the Vilsmeier reaction (7.53) (cf. reaction (7.46) for indole) and the nitration reaction (7.54) at C_2. This may be due in some way to the high electronegativity of the oxygen atom.

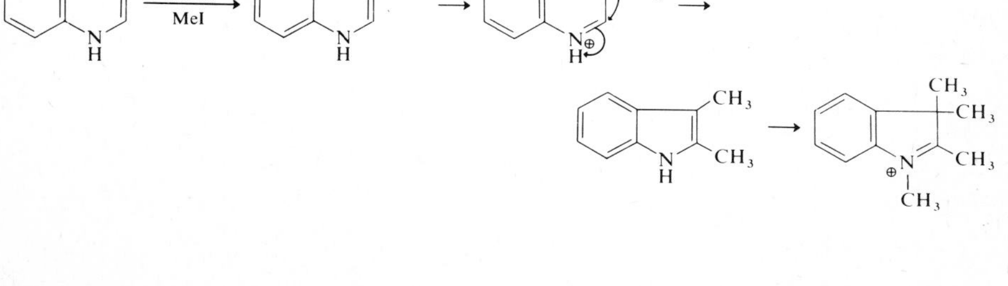

(7.52)

(7.53)

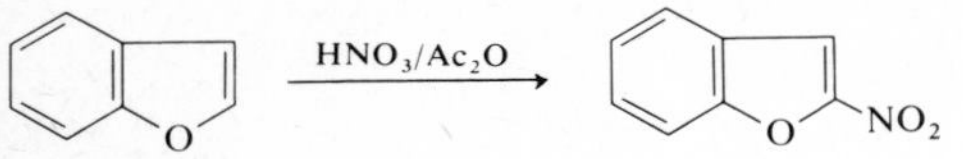

(7.54)

Metallation can be achieved with indole and benzothiophene. Indolyl salts are formed by strong bases or Grignard reagents and the N—H proton is lost to give an ambident anion (**28**) which may alkylate on nitrogen or at C_3 depending on the conditions of temperature and solvent and the nature of the positive ion and the alkylating agent. In general, the reactions favour formation of 3-substituted compounds, especially at higher temperatures.

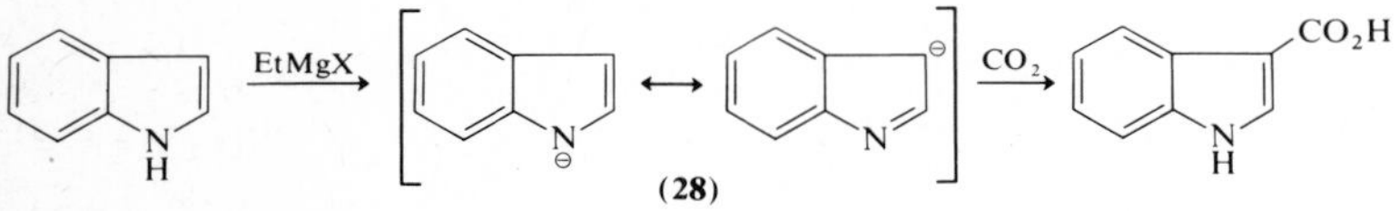

(7.55)

Benzothiophene gives sodium, lithium and bromomagnesium derivatives which react with carbon dioxide, halogens, or alkylating agents at C_2.

§ 7.2.3   Addition reactions
These can occur readily at the 2,3-double bond of indole, benzofuran or benzothiophene with electron-deficient species such as carbenes. 2,3-Dimethylindole reacts with chloroform and base as in reactions (7.56) and (7.57).

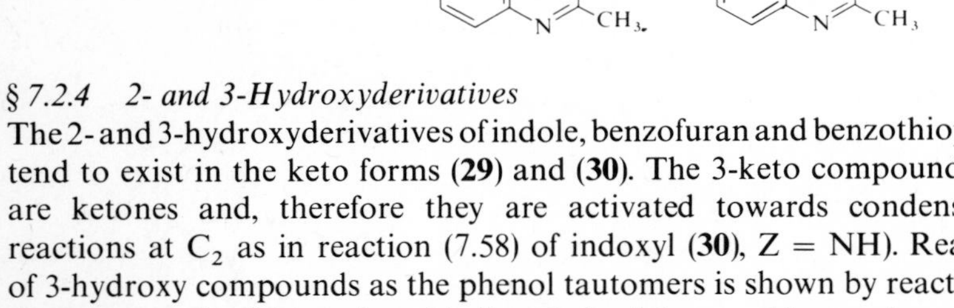

(7.56)

(7.57)

§ 7.2.4   2- and 3-Hydroxyderivatives
The 2- and 3-hydroxyderivatives of indole, benzofuran and benzothiophene tend to exist in the keto forms (**29**) and (**30**). The 3-keto compounds (**30**) are ketones and, therefore they are activated towards condensation reactions at C_2 as in reaction (7.58) of indoxyl (**30**), Z = NH). Reaction of 3-hydroxy compounds as the phenol tautomers is shown by reaction of the compounds of this series with dimethyl sulphate and base as in reaction (7.59). Alkylation with methyl iodide in nonaqueous base causes C-alkylation of the ambident anion as in (7.60). The 2-keto compounds (**29**) are typical amides, lactones and thiolactones and undergo reactions typical

of these functional groups. The indole derivative ((**29**), Z = NH) is known as oxindole. The 2,3-diones (**31**), such as isatin ((**31**), Z = NH), undergo the important reaction of condensation with a 3-keto compound as in reaction (7.61) to yield a variety of compounds belonging to the indigo series of dyestuffs (indigotin is (**32**), Z = Y = NH).

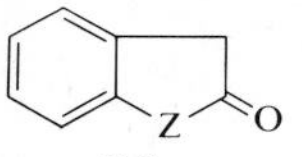

(**29**) 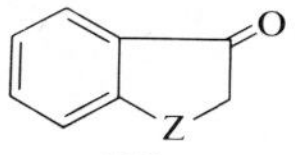   (**30**) 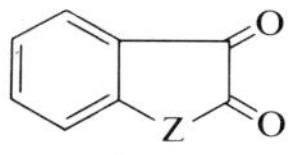   (**31**)

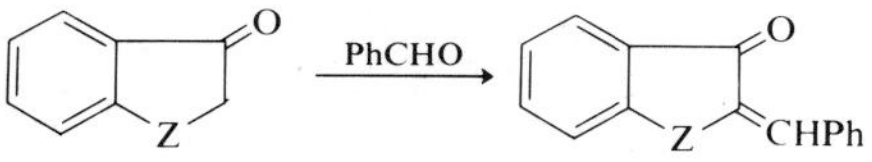

(7.58)

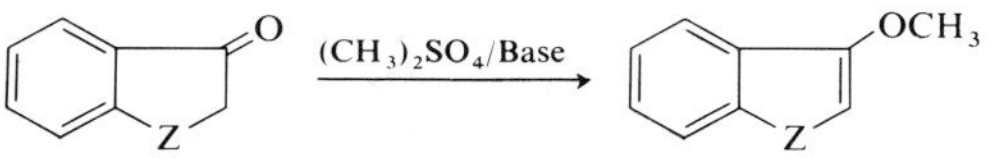

(7.59)

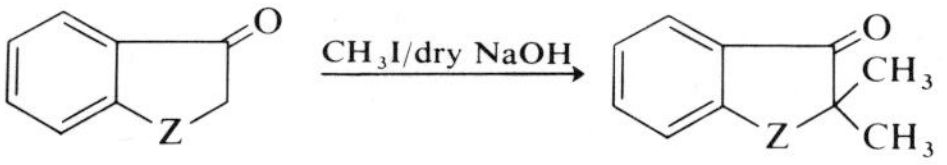

(7.60)

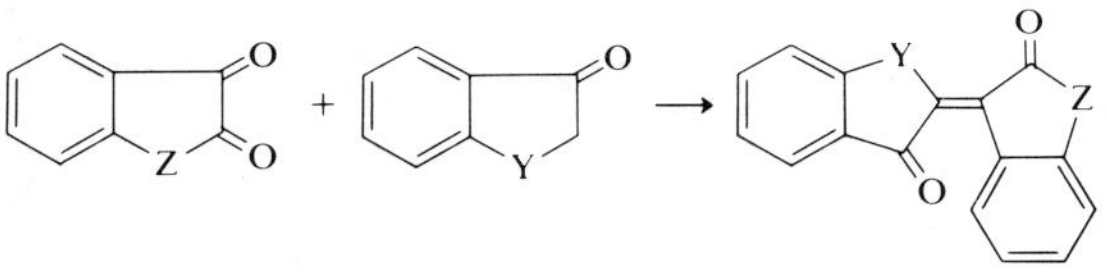

(7.61)

(**32**)

§ 7.2.5 Isoindole, isobenzofuran and isobenzothiophene

Since the compounds of this series have no Kekulé benzene ring, they are very reactive and will undergo reactions typical of dienes such as the Diels–Alder reactions (7.62). One example of a typical aromatic reaction is, however, the Mannich reaction (7.63), an electrophilic aromatic substitution reaction.

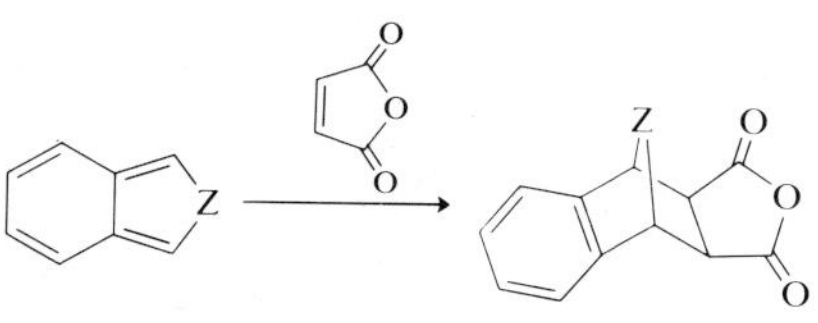

(7.62)

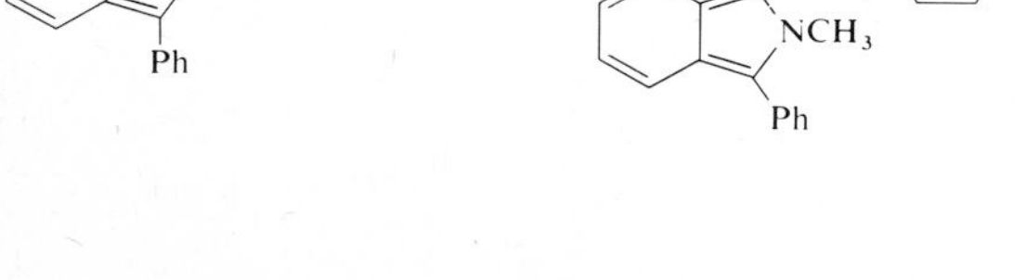

(7.63)

It has been calculated that the energy difference between isoindole (**33**) and isoindolenine (**34**) is only 33 kJ mol^{-1} and so substituents can greatly affect the tautomeric nature of any given compound.

(33)   (34)

The stability gained by the 10π-system in (**33**) seems to be balanced by the benzene sextet present in (**34**).

Isoindoles are very susceptible to oxidation and reduction reactions.

§ 7.2.6   Bicyclic compounds with five-membered rings containing more than one heteroatom

Benzimidazoles, benzoxazoles and benzthiazoles ((**35**), Z = NR, O and S) have been prepared and have many properties in common. The pyridine-type nitrogen activates methyl groups at C_2 which will react with aldehydes. This pyridine-type nitrogen also deactivates the ring towards electrophilic reagents while the other heteroatom activates the ring towards such reagents. In general, electrophilic reagents are found to attack the benzenoid ring.

Benzopyrazoles (indazoles), benzisoxazoles and benzisothiazoles ((**36**), Z = NR, O and S) are analogous compounds. Indazole is capable of existing in two tautomeric forms (**37**) and (**38**). Electrophilic reagents can substitute at C_3 in indazole and in this respect it resembles indole.

(35)   (36)   (37)   (38)

Purines are by far the most important members of this class. The parent compound (**39**) is capable of tautomerism to the 7H isomer (**40**). Since purines consist of both pyrimidine and imidazole moieties, it is not surprising that the parent compound has a base strength ($pK_a = 2\cdot5$) which is intermediate between the base strengths of pyrimidine and imidazole. It might be considered that the imidazole ring enriches the electron density of the pyrimidine ring. Purine is an unusually strong acid ($pK_a = 8\cdot9$) and this can be envisaged as being due to the anion (**41**) being resonance-stabilised as shown, with the charge distributed over four nitrogen atoms.

(39)   (40)

(41)

Alkylation occurs preferentially at N_9. As expected, positions 2, 6 and 8 are susceptible to attack by nucleophiles and 2-, 6- and 8-chloro-substituted compounds can undergo substitution of the chlorine by a variety of nucleophilic reagents in much the same way as α- and γ-chloro-pyridines can be substituted. The halogen atoms of 2,6,8-trichloropurine (42) are replaced successively in the order $6 > 2 > 8$ as shown in (7.64) below.

(42)   (7.64)

Hydroxypurines invariably exist as the amide tautomers and halogen-substituted compounds can be obtained from these by reaction with phosphorus halides.

§ 7.3   Polycyclic aromatic heterocyclic compounds

Acridine (43) and phenanthridine (44), the aza-derivatives of anthracene and phenanthrene, have many properties which can be deduced by extrapolation from the chemistry of pyridine and quinoline. Nucleophilic attack is α- or γ- to nitrogen and so nucleophilic reagents attack acridine at C_9 and phenanthridine at C_6. 9-Chloroacridine and 6-chlorophen-anthridine are readily substituted by nucleophilic reagents, and the quaternary salts of these compounds, readily obtained by alkylation, are even more reactive towards nucleophilic reagents than are the parent compounds. It is not surprising that 9-hydroxyacridine and 6-hydroxy-phenanthridine exist in the keto forms, but tautomeric behaviour has also been observed for 1- and 3-hydroxyacridines (cf. (45) $\rightleftharpoons$ (46)). 4-Hydroxy-acridine has unusual properties due to the hydrogen bonding shown in (47).

(43)   (44)   (45)   (46)   (47) H

As expected, acridine and phenanthridine react with electrophilic reagents in the benzenoid rings. Nitration of acridine occurs at C_2 and C_7. 9-Alkylacridines and 6-alkylphenanthridines are activated in much the same way as are 2- and 4-alkylpyridines (cf. § 6.6).

Alloxazines and isoalloxazines ((**48**) and (**49**) respectively) are of interest since the vitamin riboflavin contains this basic structure. These compounds may be regarded as benzopteridines and so the chemistry of the species is predictable. We shall discuss the chemistry of riboflavin in Chapter 8.

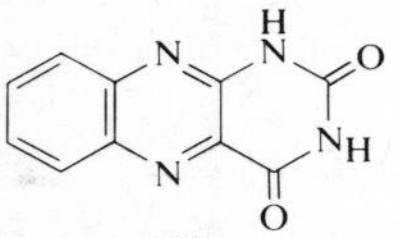

    (**48**) 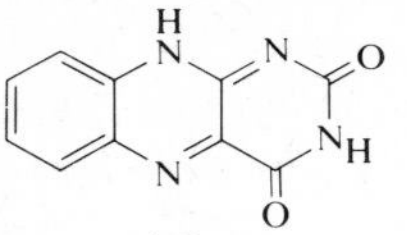   (**49**)

Carbazole (**50**) is attacked by electrophilic reagents at positions 3 and 6 and in this respect it appears to behave as a typical aniline. N-Alkylation, acylation and metallation have been observed.

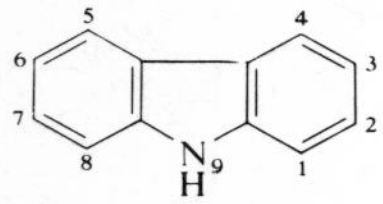

(**50**)

§ 7.4   Synthesis of polycyclic aromatic heterocyclic compounds

The most widely used synthesis of quinoline is the *Skraup* synthesis where, as shown in (7.65), a substituted vinyl ketone or aldehyde reacts with an aniline in the presence of an oxidant to yield a quinoline. Anilines can also be reacted with 1,3-dicarbonyl compounds, as in reactions (7.66) and (7.67), to yield substituted quinolines and this reaction can be adapted to phenols to make benzopyrilium salts or (as in (7.68)), coumarins. It is best to use symmetrically substituted 1,3-dicarbonyl compounds in these reactions if mixtures are not to result.

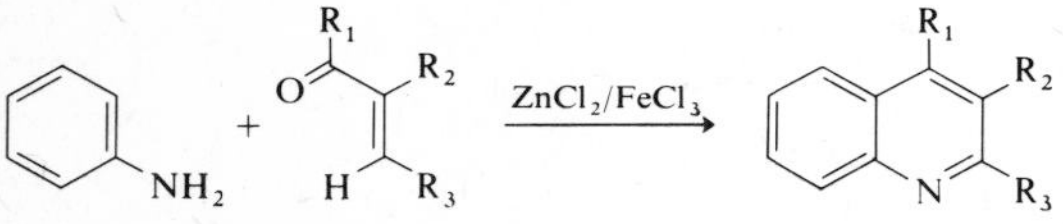

(7.65)

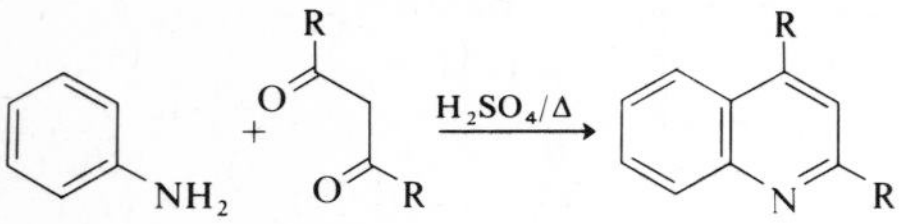

(7.66)

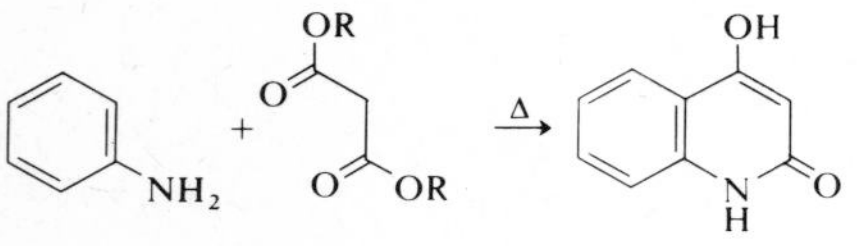

(7.67)

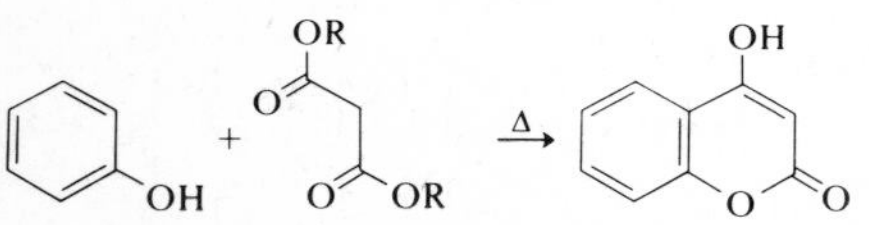

(7.68)

In the *Friedlander* synthesis of quinoline (7.69), *ortho*-acyl anilines react with ketones with a free α-methylene group to yield quinolines. In the *Pfitzinger* modification of this (reaction (7.70)), isatin is used instead of an *ortho*-aminobenzaldehyde and the resultant carboxylic acid can be very readily decarboxylated. The *Friedlander* synthesis also has its counterpart (*e.g.* (7.71)) in the synthesis of coumarins.

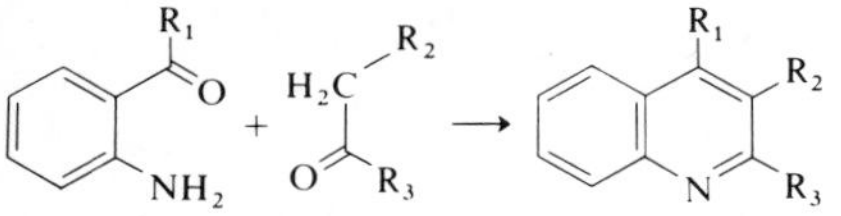

$$(7.69)$$

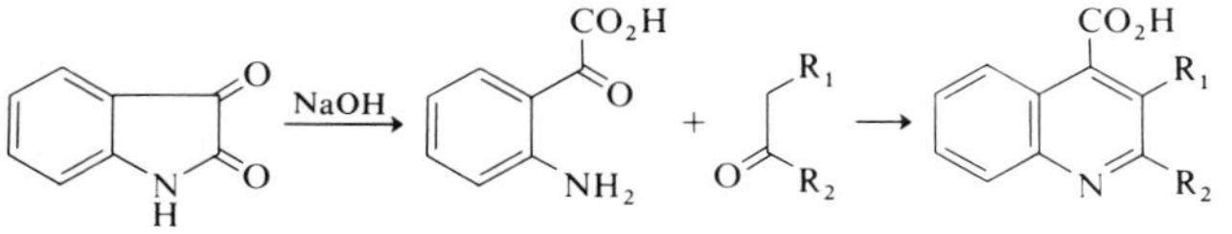

$$(7.70)$$

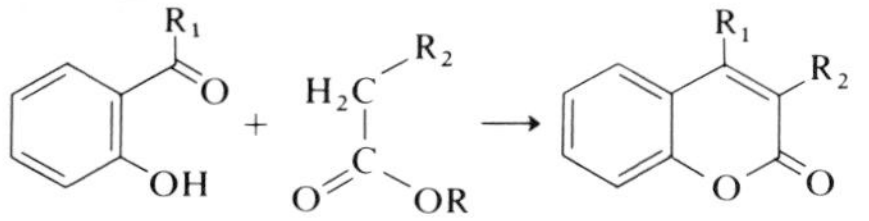

$$(7.71)$$

Two commonly used syntheses of isoquinoline are the *Bischler–Napieralski* (7.72) and the *Pictet–Spengler* (7.73) syntheses. Both of these use 2-phenylethylamine as their starting point, and both require the use of an oxidant in a final step.

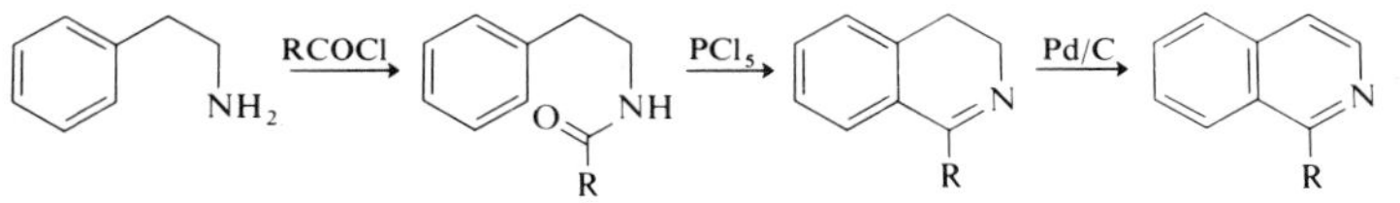

$$(7.72)$$

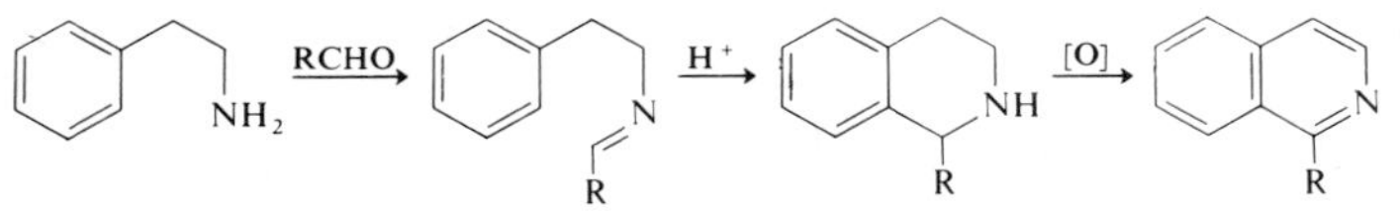

$$(7.73)$$

There are several general methods for making other heterocyclic compounds. Reaction of *ortho*-diamines with α-dicarbonyl compounds has been used in synthesis of quinoxalines (7.74), pteridines (7.75) and alloxazines (7.76), while reaction of formic acid with diamino compounds has been effective in synthesis of quinazolines as in (7.77) and of purines as in (7.78).

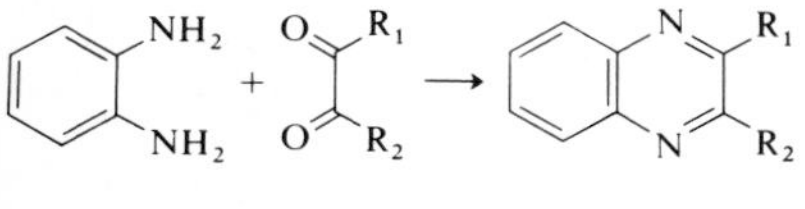

$$(7.74)$$

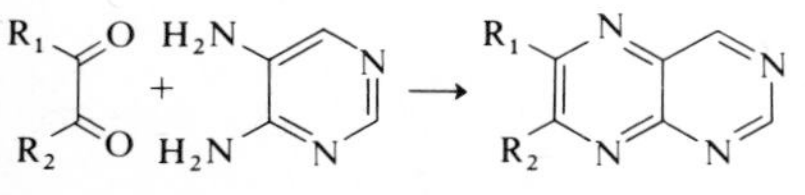

$$(7.75)$$

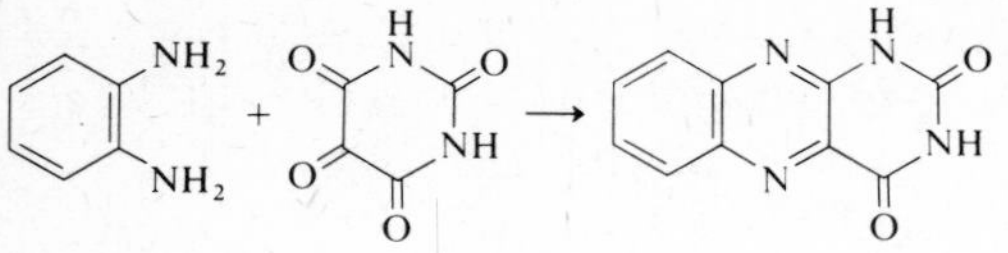

(7.76)

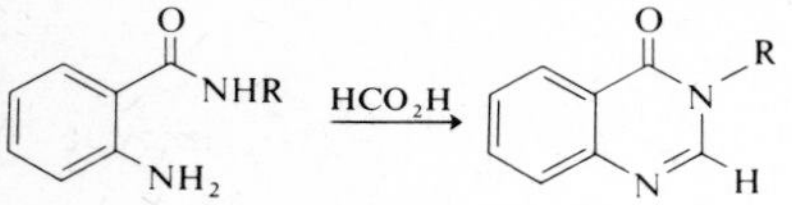

(7.77)

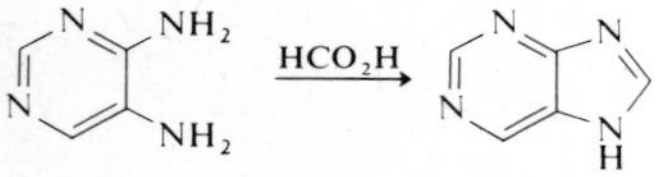

(7.78)

The most widely used synthesis of indoles is the *Fischer Indole* synthesis outlined in (7.79) below. This involves rearrangement of the phenyl-hydrazone of a ketone or aldehyde by heating it in the presence of anhydrous zinc chloride.

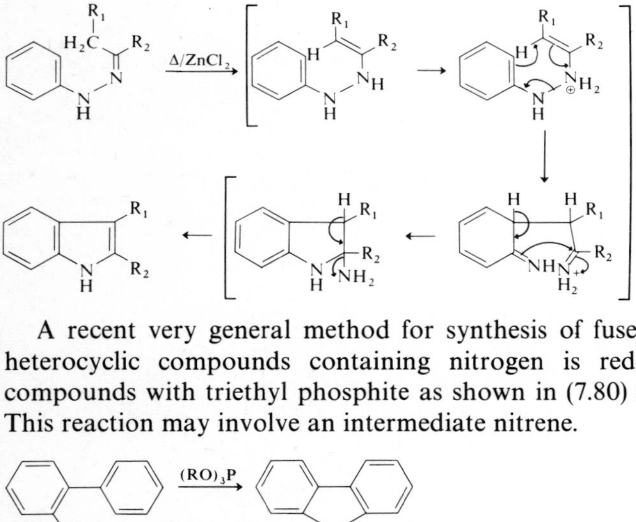

(7.79)

A recent very general method for synthesis of fused five-membered heterocyclic compounds containing nitrogen is reduction of nitro-compounds with triethyl phosphite as shown in (7.80) and (7.81) below. This reaction may involve an intermediate nitrene.

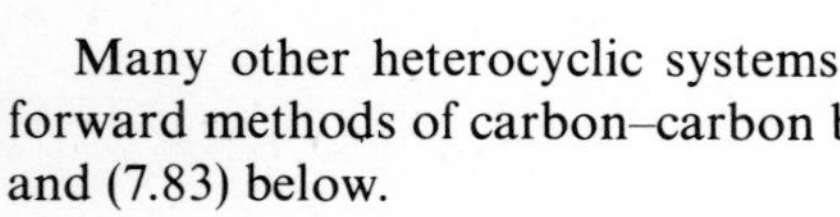

(7.80)

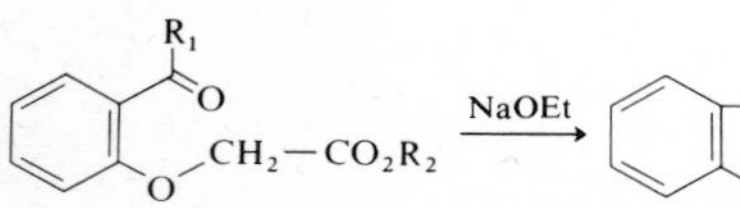

(7.81)

Many other heterocyclic systems have been synthesised by straight-forward methods of carbon–carbon bond formation as in reactions (7.82) and (7.83) below.

(7.82)

(7.83)

Key points

(1) Although polycyclic heteroaromatic compounds have properties which may be predicted from the chemistry of the monocyclic "parent" compounds, there are also properties which are peculiar to the fused compounds themselves.

(2) When six-membered heterocyclic and benzenoid rings are fused, then it is to be expected that the heterocyclic ring will be less prone to attack by electrophilic reagents and more prone to attack by nucleophilic reagents than the benzene ring.

(3) The effects of quaternisation and N-oxide formation are predictable from the chemistry of the monocyclic systems. Chloro- and alkyl-substituted compounds also react in a predictable manner.

(4) Fused bicyclic compounds with a five-membered heteroaromatic ring can have properties which are very different from the monocyclic parent compounds. Because of the relative stability of the benzene ring, indole is substituted by electrophilic reagents at C_3 whereas pyrrole is substituted at C_2. For the same reason, benzothiophene sulphone is less like a diene in its reactions than is thiophene sulphone.

(5) Compounds such as isoindole are 10π-electron systems, but since they lack a benzene ring and reaction with dienophiles will generate such a ring, they behave like dienes.

Further reading

General

R. C. Elderfield, ed. *Heterocyclic Compounds*, vols. 2, 3 and 4, Wiley, New York, 1951, 1952, 1952.

Pteridines

W. Pfleiderer, *Angewandte Chemie International Edition*, 1964, **3**, 114.

Indole

W. J. Houlihan, ed. *Indoles*, Parts 1 and 2, Wiley-Interscience, New York, 1972.

R. J. Sundberg, *The Chemistry of Indoles*, Academic Press, New York, 1970.

Isoindoles

J. D. White and M. E. Mann, *Advances in Heterocyclic Chemistry*, 1969, **10**, 113, Academic Press, New York.

Acridines

R. M. Acheson, ed. *Acridines*, second edition, Wiley-Interscience, New York, 1973.

Chapter 8

Heterocyclic compounds in nature and medicine

Heterocyclic systems occur in a wide variety of natural and non-natural compounds. A knowledge of heterocyclic chemistry can, therefore, be useful in the areas of natural product chemistry and biosynthesis and in studies of drug metabolism. A large number of coenzymes, vitamins and drugs rely on heterocyclic reactivity for their action and we shall see in this chapter how the principles of heterocyclic chemistry outlined in the previous chapters can help in the understanding of the mode of action of such compounds.

§ 8.1   Heterocyclic compounds which mediate in biochemical processes

§ 8.1.1   Nucleic acids

Purines and pyrimidines are heterocyclic systems which are important in the most basic biological processes of heredity and evolution. The genetic material, deoxyribonucleic acid (DNA), has been shown to consist of a polymeric sugar phosphate chain as in (6) below. Deoxyribose is the monomeric sugar unit to which are attached at C_1, a purine, adenine (1) or guanine (2), or a pyrimidine, thymine (3) or cytosine (4). A second type of nucleic acid which is responsible for conveying the genetic message from DNA is ribonucleic acid (RNA). This is similar in structure to DNA but the sugar monomer is ribose, and uracil (5) replaces thymine (3) as one of the four bases attached to C_1 of the ribose units. DNA has been

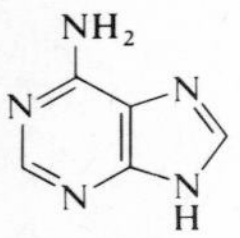

(1) 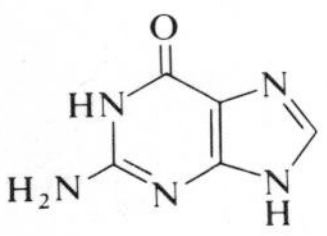   (2) 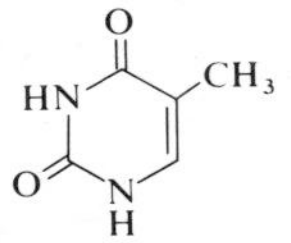   (3) 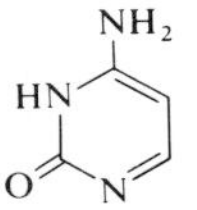   (4) 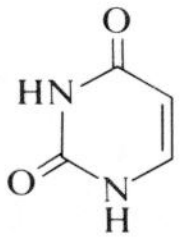   (5)

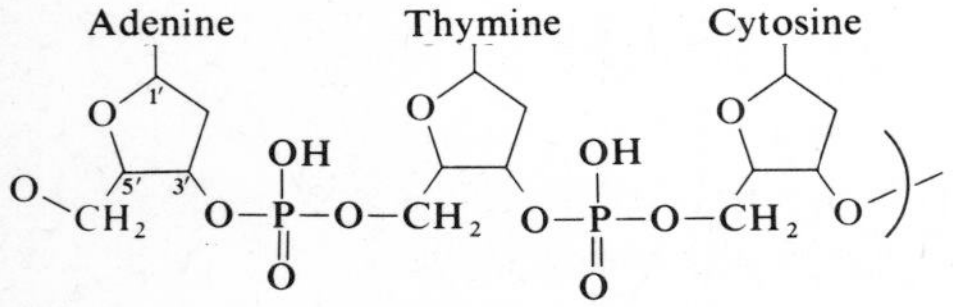

(6)

110

shown to consist of two helical interacting strands which split into individual strands on cell division. These strands then synthesise matching strands in the new cells. The individual strands contain the basic information of the genes and so, for replication, the initial strand must synthesise its matching strand in one way and one way only. The way in which this replication is achieved was discovered when it was realised what the correct tautomeric forms of the individual purine and pyrimidine bases were. From the discussion in Chapter 4 (§ 4.4), the tautomers will be those described in formulae (1) to (5), and when models of these are made it can be seen that strong hydrogen bonding is possible between thymine (or uracil) and adenine and between cytosine and guanine as in (7) and (8) below, and that the distance "A" in each of these pairs is the same. This means that strands can only be made up by the base-pairing (7) or (8), and so in DNA strand A can form only one complementary strand, strand B as shown in (9).

Thymine (R = CH$_3$)   Adenine   Cytosine   Guanine
Uracil (R = H)
(7)   (8)

A = Adenine
G = Guanine
C = Cytosine
T = Thymine

(9)

The replication of genetic material is, therefore, intimately related to heterocyclic structure. DNA is indirectly responsible for synthesis of the protein enzymes via messenger RNA and the unravelling of the genetic code with its "alphabet" of four letters, the two purines and the two pyrimidines, and its three letter "words" to code for each of the twenty possible amino acids, is one of the more fascinating pieces of recent biological research.

§ 8.1.2   *Other compounds involving purines and pyrimidines*
The purine adenine occurs in a very large number of biologically useful compounds. The widespread occurrence of such compounds may be due to the very easy synthesis of adenine from ammoniacal hydrogen cyanide. This could mean that before life was present on earth adenine could have been present in abundance and so might have been incorporated into living systems. The scheme for synthesis of adenine from hydrogen cyanide is outlined in (8.1) below. Some of the large variety of biologically useful

compounds containing adenine are coenzyme A **(10)** which is responsible for the transfer of acyl groups in nature; S-adenosyl-L-methionine **(11)** which is involved in methyl transfer reactions in nature; the redox coenzyme NAD **(12)**; and adenosine triphosphate and a variety of other compounds.

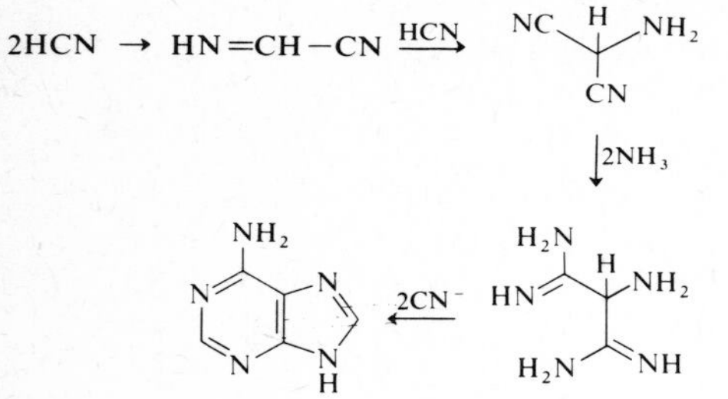

$$2HCN \rightarrow HN{=}CH{-}CN \xrightarrow{HCN} \underset{CN}{\overset{NC}{\underset{}{\diagdown}}} \overset{H}{\underset{}{C}} \diagup NH_2$$

(8.1)

$$\downarrow 2NH_3$$

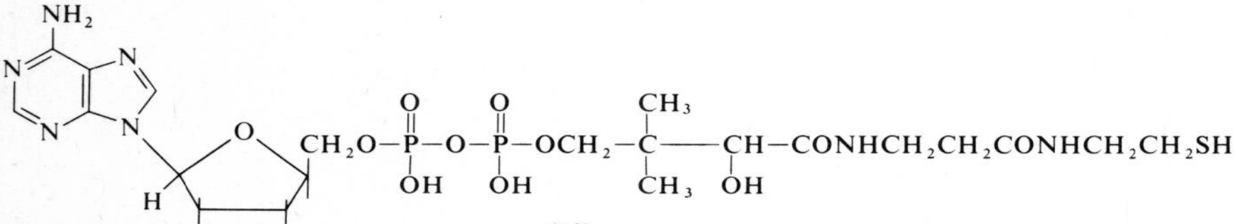

(10)

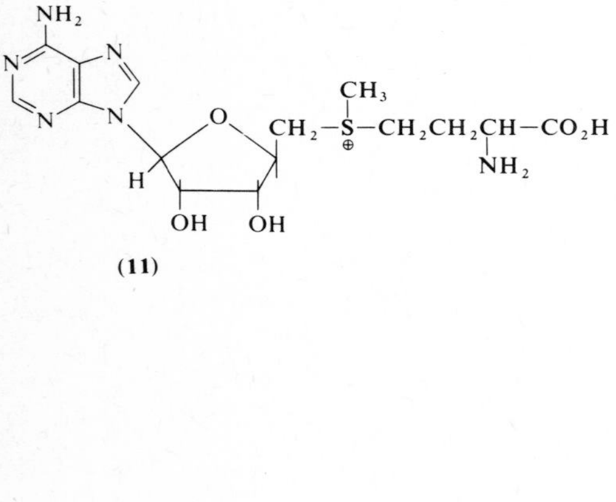

(11)

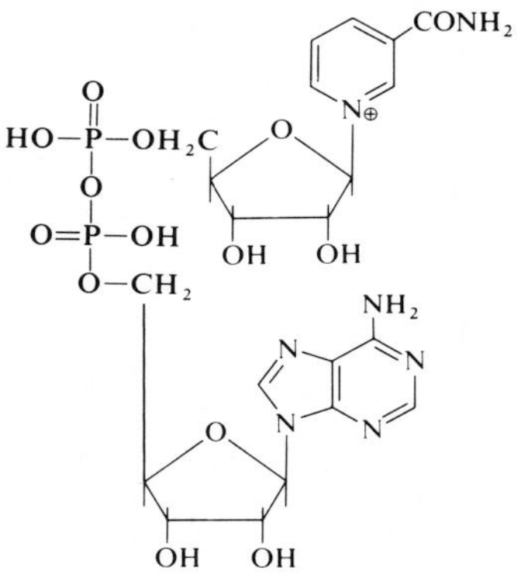

(12)

Since DNA is involved in cell division and since tumour cells divide more rapidly than healthy cells, various purine and pyrimidine derivatives such as **(13)** and **(14)** are useful as tumour inhibitors. This is because these compounds are structurally similar to the DNA bases and will interfere with DNA synthesis. Bacterial cells divide more rapidly than the cells of their host organism, and so some purines and pyrimidines can, by interference with DNA synthesis, act as antibiotics. Puromycin **(15)** is an example of such an antibiotic.

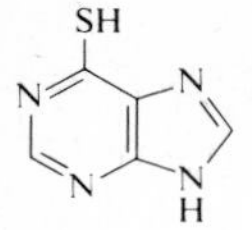

(13) 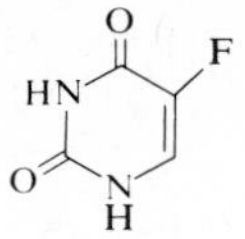   (14) 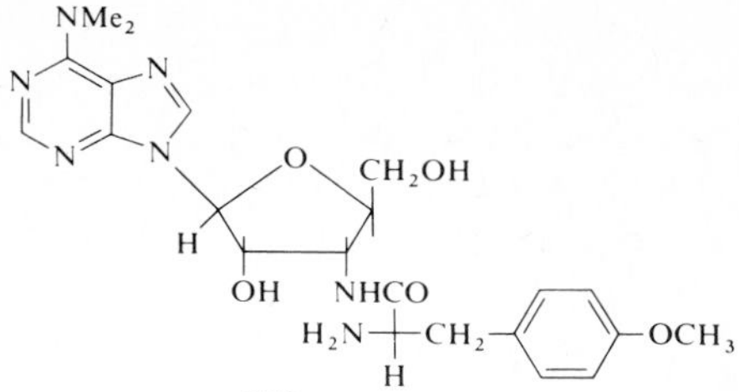   (15)

§ 8.1.3   Imidazoles as biologically useful compounds

Chymotrypsin is the enzyme whose mechanism has been most thoroughly studied. It is responsible for the transfer of the acyl group from a number of donors such as amides, acids and esters to a number of acceptors such as water, alcohols and amines. In its function as the human digestive enzyme, it acts as a hydrolase and so the acceptor is water. The polypeptide enzyme has two important features in its active site, the hydroxymethyl moiety of the amino acid serine (16) and the imidazole moiety of the amino acid histidine (17). One model for the function of chymotrypsin as an ester hydrolase is outlined in (8.2) below. Here the hydroxymethyl function is responsible for ester formation while imidazole is required because of its acid-base properties (discussed in § 5.2.1 and § 5.2.3).

(16)   (17)

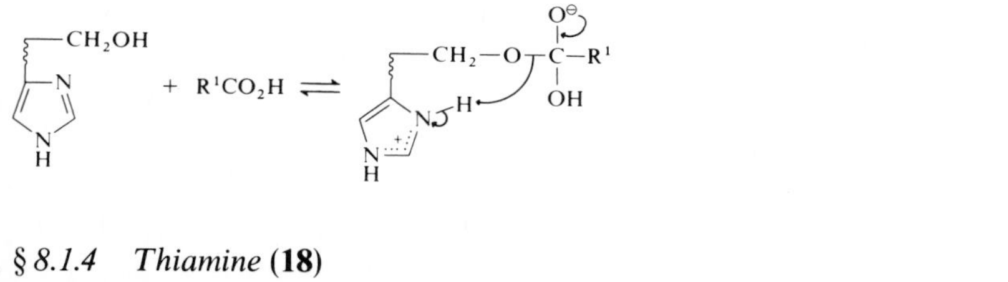

(8.2)

§ 8.1.4   Thiamine (18)

Unlike chymotrypsin, which is an enzyme which requires no coenzyme or cofactor, many enzymes operate through the intermediacy of small molecules termed coenzymes or cofactors. These mediate in a variety of biological reactions by virtue of their chemical structure. Thiamine (18) is one such coenzyme and has a structure which contains both a pyrimidine and a thiazolium ring. We have seen in § 5.5 that thiazolium salts have a very acidic hydrogen at C_2 and readily form zwitterions such as **(19)**. Thiamine operates as a coenzyme in reactions which, depending on the enzyme, can be decarboxylations as in (8.3), acetoin condensations such as (8.4) and transketolase reactions such as (8.5). In all of these reactions the coenzyme reacts by virtue of the nucleophilicity of the zwitterion **(19)**, the strongly electrophilic nitrogen in intermediates such as **(20)**, **(21)** and **(22)** and the leaving group properties of the zwitterion.

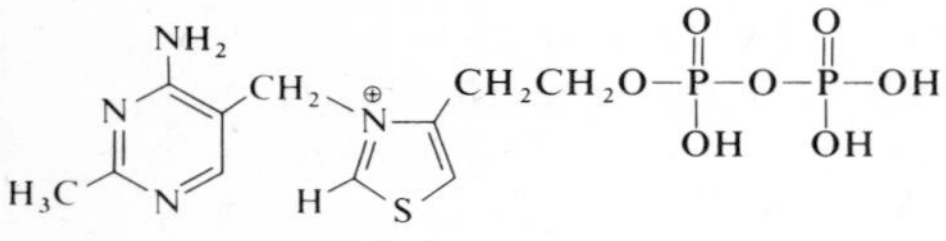

(18) 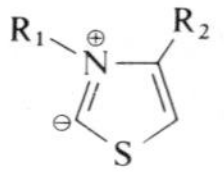   (19)

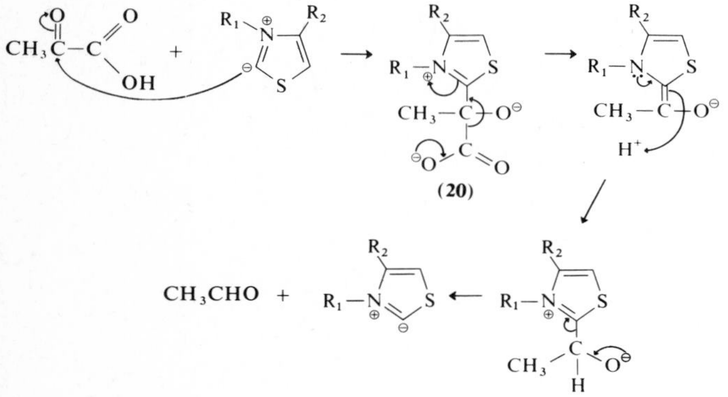

(20)

$$CH_3CHO \; + \; \text{thiazolium} \; \longleftarrow \; \text{thiazolium} \qquad (8.3)$$

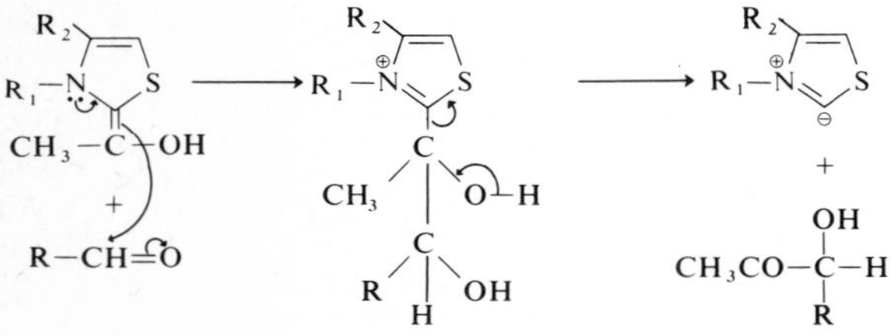

(21)

$$\qquad (8.4)$$

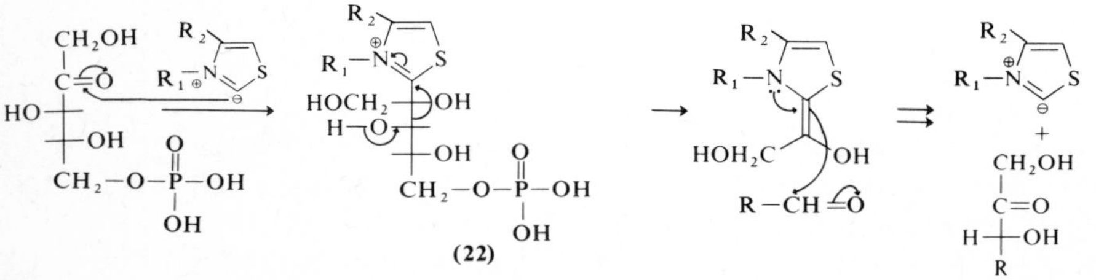

(22)

$$\qquad (8.5)$$

§ 8.1.5 Pyridine in biological processes

Nicotine adenine dinucleotide (NAD (**12**)) is an important redox coenzyme which owes its reactivity to the pyridinium moiety in the molecule. We have seen how quaternary pyridinium salts (§ 6.5) are very susceptible to attack by nucleophilic reagents at positions 2, 4 and 6. The pyridinium moiety will accept "hydride" from an alcohol to yield the 4(H) derivative and a ketone or aldehyde. This may be achieved by initial attack at either C_4 or C_2 as shown in reactions (8.6) and (8.7) respectively. The aromaticity of the pyridinium salt makes the reverse reaction readily undergone.

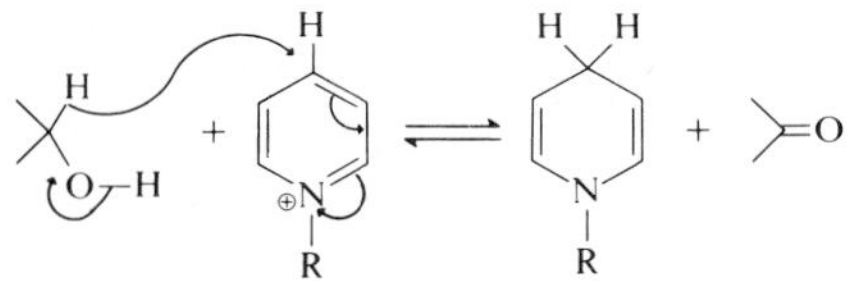

(8.6)

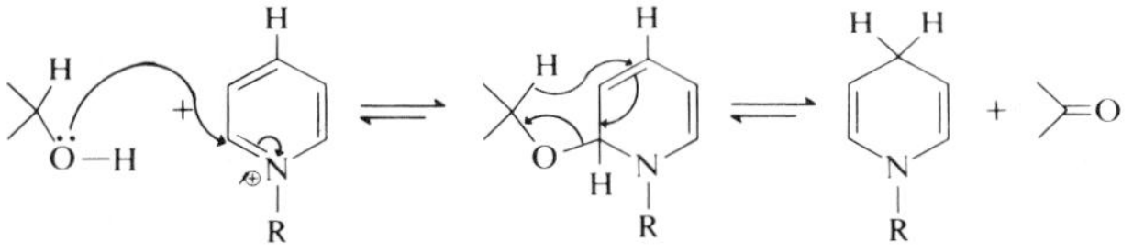

(8.7)

Pyridoxal phosphate is a cofactor in a variety of important biological processes such as transamination, racemisation and various elimination reactions. The cofactor, a pyridine aldehyde, gives a very electrophilic pyridinium nitrogen on protonation and this aids rearrangement of the intermediate Schiff's base (*e.g.* (**23**)) in transamination reactions (8.8) and in tryptophan synthesis (8.9).

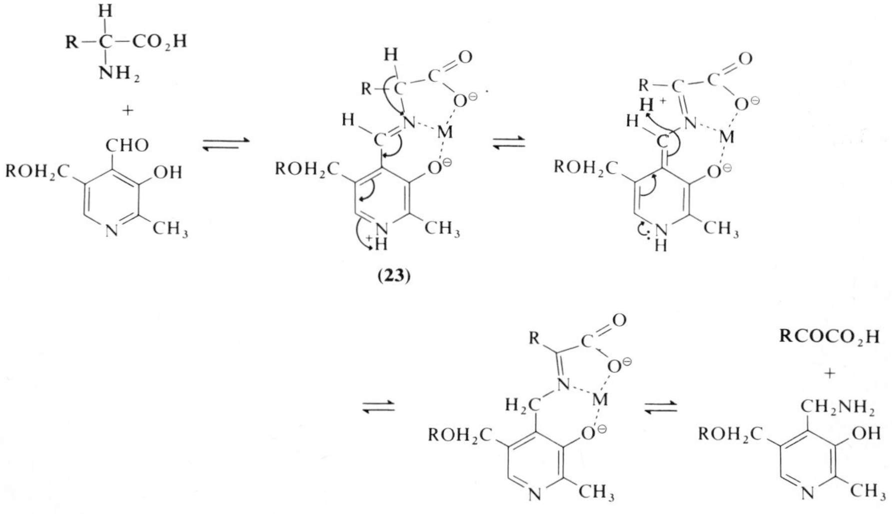

(**23**)

(8.8)

(8.9)

§ 8.1.6   *Flavin coenzymes*

Riboflavin, the human nutritional factor vitamin B_2, was found to have the isoalloxazine structure (**24**) and was recognised to be an important coenzyme in a variety of biological dehydrogenation and oxygen-transfer reactions. A knowledge of heterocyclic chemistry has allowed for speculation on the mechanism of these reactions and has suggested experiments which may eventually lead to a true understanding of the action of the coenzyme.

It is known that the reduced compound (**25**) is involved in riboflavin-mediated oxygen transfer reactions. In (**25**) the system $N_5-C_{4a}-C_{10a}$ is an enamine while $N_{10}-C_{10a}-C_{4a}$ and $N_1-C_{10a}-C_{4a}$ are conjugated to the C_4 carbonyl group and so are vinylogous amides. It is, therefore, possible that autoxidation of (**25**) is similar to autoxidation of indole (§ 7.2.2) in occurring at the electron-rich carbon, C_{10a}. A reasonable mechanism for biological oxygen-transfer reactions is outlined in (8.10) below. Since alloxazines and isoalloxazines may be regarded as benzo-pteridines, it is not surprising that reduced pteridines such as the compound (**26**) are known to act as coenzymes for hydroxylases and the mechanism (8.11) would be analogous to the mechanism (8.10) of flavin-mediated oxygen transfer reactions.

In biological dehydrogenation reactions such as the dehydrogenation of alcohols to ketones, the flavin coenzyme undergoes the redox reaction (**24**) $\rightleftharpoons$ (**25**) and the reactions are known to be reversible. It has been suggested that, since in (**24**) the imine carbon 4a will be very electrophilic, the mechanism outlined in (8.12) below may account for these reactions.

If the scheme is to be reversible, however, then the reverse reaction would require nucleophilic attack by the flavin *at the carbonyl oxygen* of the substrate. There is no precedent for a reaction of this sort.

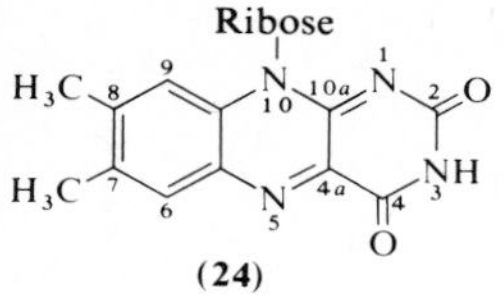

(24) 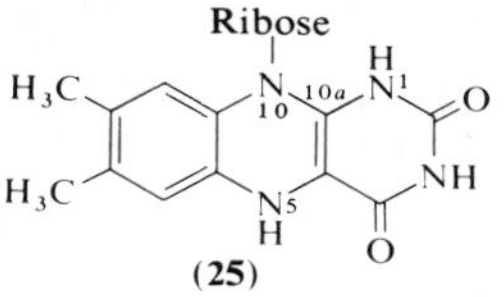 (25)

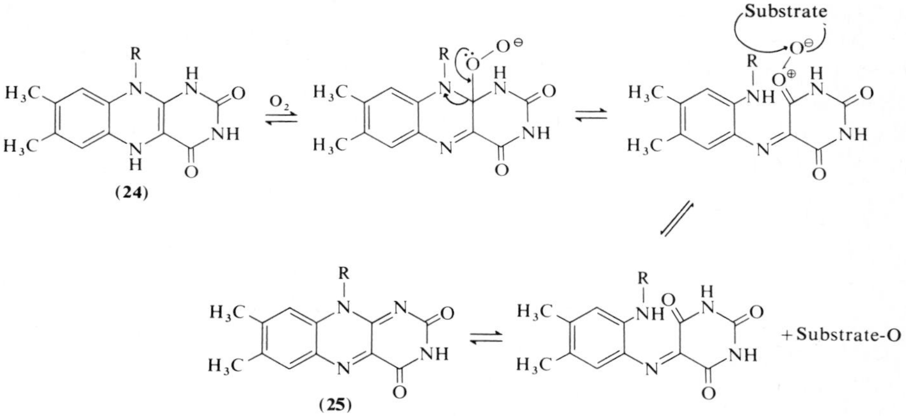

(24) $\rightleftharpoons$ $\rightleftharpoons$ Substrate

$\rightleftharpoons$ + Substrate-O (8.10)

(25)

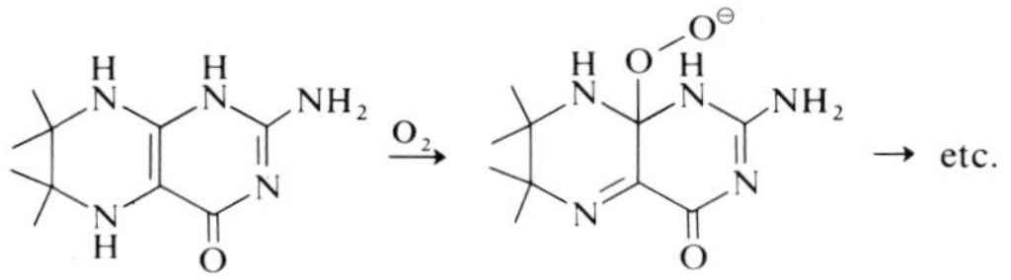

(26) $\xrightarrow{O_2}$ $\rightarrow$ etc. (8.11)

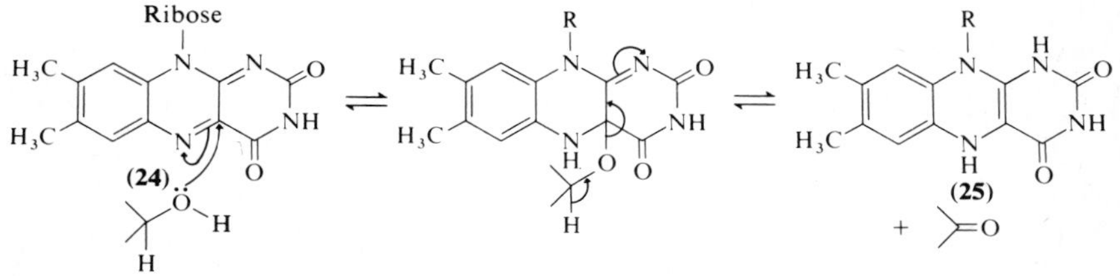

(24) $\rightleftharpoons$ $\rightleftharpoons$ (25) + (8.12)

§8.1.7 *Aliphatic heterocyclic compounds as coenzymes*

Biotin, the human growth factor vitamin H, was found to have the structure (27) and it has been shown that biotin is a carboxyl-transfer coenzyme which acts via the N-carboxy compound (28). The coenzyme lipoic acid, originally isolated from beef liver, has the structure (29) and is involved in acyl transfer reactions. The facile cleavage of the disulphide link affords a redox system.

(27)   (28)   (29)

§8.1.8 Porphyrins and corrins

The porphyrin nucleus (30) can be made by fusing four pyrrole rings together in a macrocyclic ring. Such compounds may be thought of as being 18π aromatic systems having nine double bonds in closed planar cyclic conjugation from C_1 to C_{18}. Two double bonds "ab" and "cd" are not involved in this aromatic system, and these might be expected to be olefinic and therefore more reactive. The porphyrin nucleus can react with a variety of metals including magnesium, iron, zinc, nickel, cobalt, copper and silver to give compounds of general structure (30).

In *haem*, the oxygen-transfer coenzyme, ferrous ion is bound to the porphyrin moiety as is "M" in (30) and the fifth and sixth ligand sites of the ferrous ion are occupied by imidazole nitrogen from protein histidine. Oxygen may replace one of these imidazole moieties to form an oxygen complex which can be transported in the blood to where it is required by the body. Iron remains in the ferrous oxidation state throughout the process.

Cytochromes are present in the cells of aerobic organisms and contain porphyrin groups with iron as the complexed metal. The potential (E_0) at which the redox reaction, $Fe^{2+} \rightleftharpoons Fe^{3+} + e$, occurs in cytochromes is dependent on the precise nature of the ligands round iron and there is a series of cytochromes differing in this potential by very small amounts. This allows electron transport to occur in a series of closely related cytochromes, the first and last redox steps having quite widely different potentials. The terminal member of the cytochrome chain has an e.m.f. close to that required for the redox reaction, $\frac{1}{2}O_2 + H_2 \rightleftharpoons H_2O$, and the chain of cytochromes may use this reaction to achieve redox reactions at quite different potentials from that of the terminal reaction.

(30)   (31)   (32)

In the photosynthetic pigment chlorophyll (31) one of the "non-aromatic" double bonds "ab" and "cd" of the porphyrin is reduced, and a bacterial chlorophyll is known in which both of these double bonds are reduced. The important compound vitamin B_{12} has the closely related corrin nucleus (32).

§8.1.9   Bioluminescent compounds

Cypridina luciferin (33) is the active principle in a luminescent crustacean and reacts like a conjugated enamine with oxygen as in (8.13) below. The resultant peroxy anion can react as a nucleophile with the carbonyl group of the imidazolone in (34) and the attack results in a dioxetane (35). The reaction is thermodynamically favoured by the formation of the aromatic pyrazine ring in (35). The unstable dioxetane (35) will break up by a "forbidden" process as shown, resulting in emission of energy.

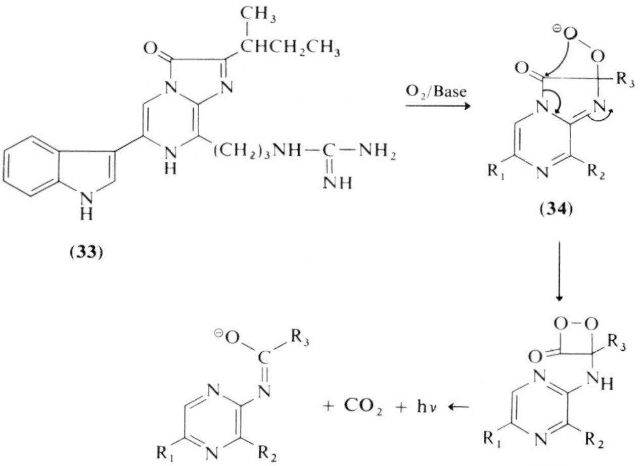

(8.13)

§8.2   Heterocyclic compounds as drugs

A large number of pharmaceutically useful compounds are heterocyclic compounds. A considerable amount of research has been done to deduce some relationship between the structure of a compound and its activity and, although many clues to such relationships have been investigated, the field is not well understood.

In Chapters 2 and 3 it has been shown that small ring aliphatic heterocyclic compounds are very reactive and there is a large number of drugs which seem to owe their activity to the reactivity of small heterocyclic rings.

Aziridine drugs such as the antibiotic *mitomycin C* (36) and the anti-cancer nitrogen mustard drugs of the *tetramine* (37) type owe their activity to the fact that aziridine rings can undergo ring-opening reactions with nucleophiles (cf. § 2.6) as in reaction (8.14). Thus the nitrogen mustards can cross-link two strands of the helix of DNA by alkylating nueleophilic groups on the purine and pyrimidine bases and so interrupt cell division. This makes these compounds useful anti-cancer drugs. Since bacteria undergo cell division more rapidly than their hosts, mitomycin C (**36**) may act as an antibiotic by cross-linking in a similar way to the nitrogen mustards; antibiotics of this type are being screened for possible uses in cancer therapy.

(36)   (37)

(8.14)

In § 3.3.1 and § 3.3.2 we have seen that three- and four-membered ring lactams are very much more reactive than are acyclic amides or larger-ring lactams. The additional ring strain of resonance form (**38b**) compared to resonance form (**38a**) makes the carbonyl group of β-lactams more like an ester carbonyl and this is reflected in carbonyl absorption in the infra-red at $1730 \, \text{cm}^{-1}$ compared to the absorption of acyclic amides at $1640 \, \text{cm}^{-1}$. The preponderance of resonance form (**38a**) makes the β-lactam carbonyl more electrophilic than an acyclic amide, and, after addition of the nucleophile, ring-strain makes for easy ring opening, as we have seen in § 3.3.2. The antibiotics *penicillin* (**39**) and *cephalosporin* (**40**) are β-lactams and fusion of the second heterocyclic ring to the β-lactam ring has the effect of moving the amide carbonyl absorption to $1770 \, \text{cm}^{-1}$, implying that there is an even more reactive carbonyl group present. This may be due to the fact that the ring fusion does not allow the groups round the amide nitrogen to achieve the planarity required in the sp^2 hybridised resonance form (**38b**). The resultant preponderance of form (**38a**) will make for even greater electrophilicity than is present in simple β-lactams. It is thought that the high reactivity of the β-lactam is essential to the antibiotic activity of these compounds. Penicillins and cephalosporins act by inhibiting the terminal step of bacterial cell wall synthesis which involves cross-linking of peptidoglycan strands. The antibiotics could interfere with this by acylating free thiol groups as in reaction (8.15).

(38a)   (38b)

(39)   (40)

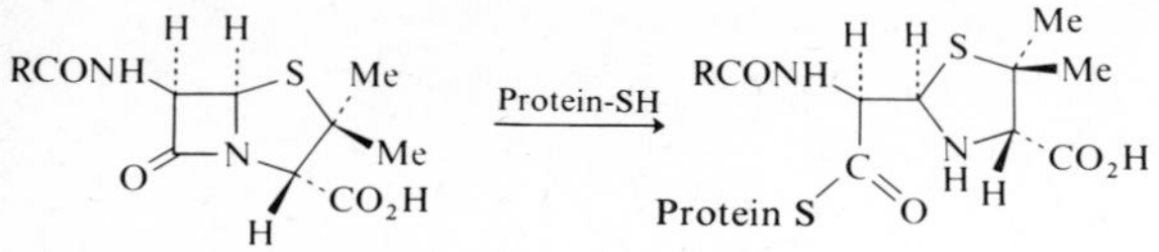

$$(8.15)$$

The alkaloid cocaine (**41**) is a bicyclic heterocyclic compound with a very rigid conformation. It was at one time a very useful local anaesthetic. The spatial arrangement of the amine and the benzyl ester groups is important to the action of this drug and the conformationally mobile procaine (**42**) which can adapt better to the drug receptor site is an important local anaesthetic. Histamine (**43**) is a compound whose release is responsible for many allergies and analogues have been prepared which have antihistamine properties.

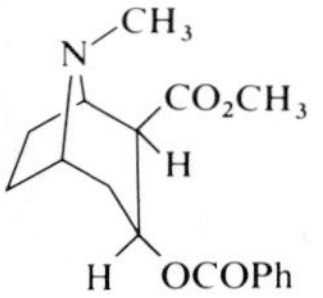

(**41**) 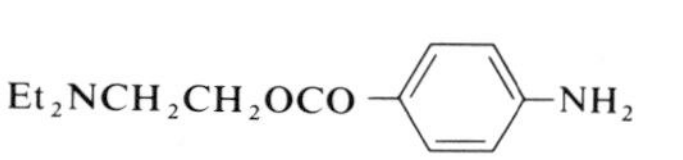 (**42**) 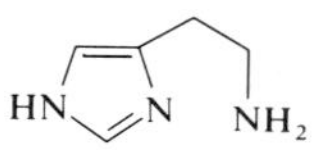 (**43**)

The narcotic nicotine (**44**) has a pyrrolidine and a pyridine ring and quaternised pyridines such as diquat (**45**) and paraquat (**46**) are important herbicides, acting by interference with photosynthesis, presumably due to the redox properties of pyridinium salts (§ 6.5). The triazine simazine (**47**) and 3-aminotriazole (**48**) also interfere with photosynthetic processes.

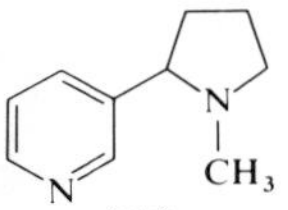

(**44**) 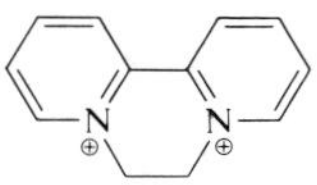 (**45**) 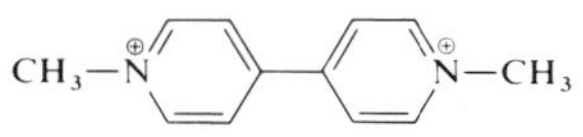 (**46**) 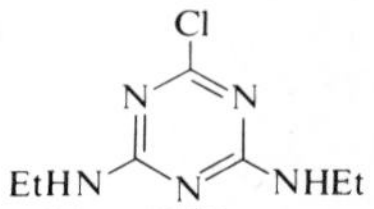 (**47**)

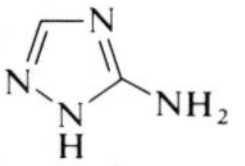

(**48**)

Quinine (**49**), a naturally occurring antimalarial drug, has both quinuclidine and quinoline rings and synthetic antimalarials having quinoline rings have been prepared. 8-Hydroxyquinoline prevents bacterial growth by its ability to complex with metal ions, as illustrated in (**50**). Coumarins abound in nature, and the finding that dicoumarol (**51**), found in sweet clover, was an anticoagulant led to the discovery that other coumarins had this property and could be used as rat poisons.

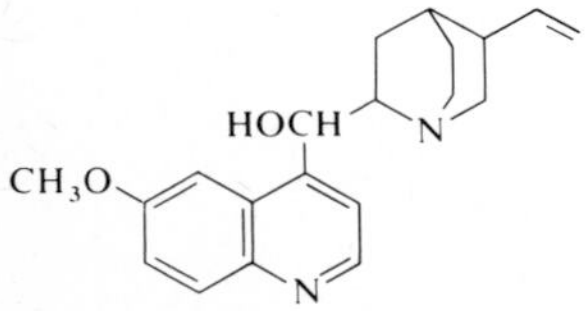

(**49**) 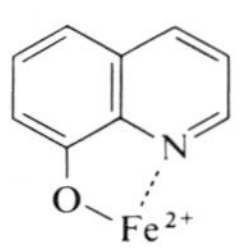 (**50**) 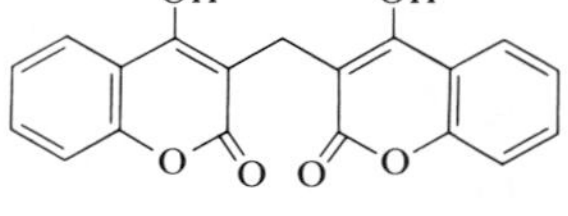 (**51**)

Barbiturates such as phenobarbitone (52) and a variety of tranquillising drugs such as imipramine (53), largactil (54) and librium (55) have heterocyclic structures. There is a large variety of indole, quinoline and isoquinoline alkaloids which have drug action and examples of these are the smooth-muscle relaxant papaverine (56), the amoebicide emetine (57) and the hallucinatory drug LSD (58).

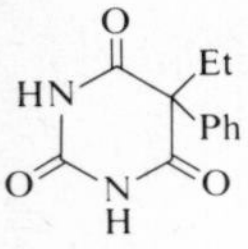

(52) 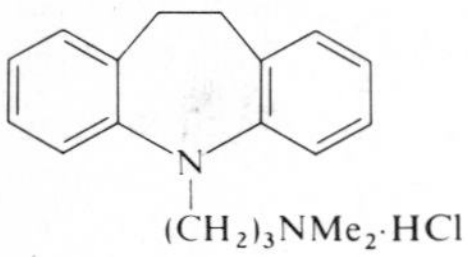   (53) 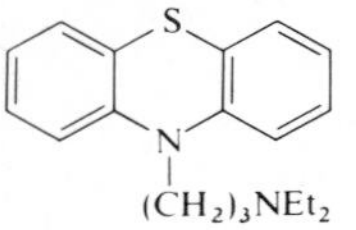   (54) 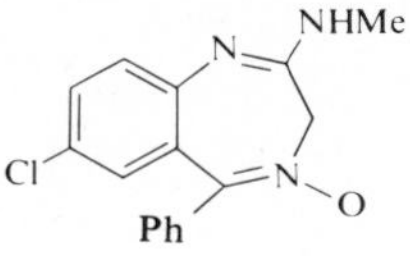   (55)

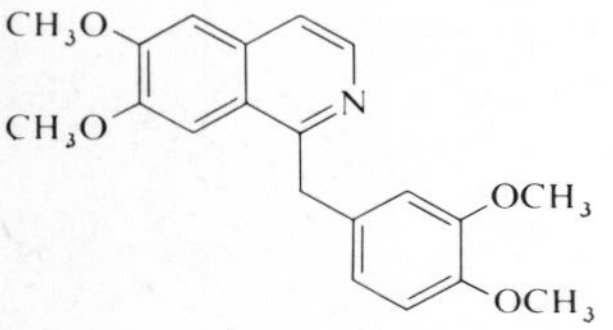

(56) 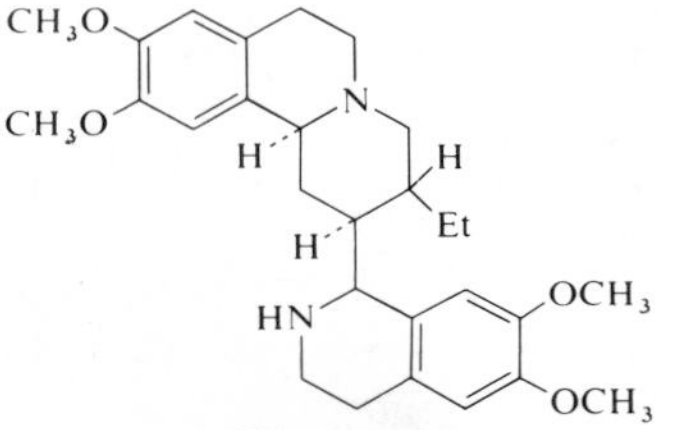   (57) 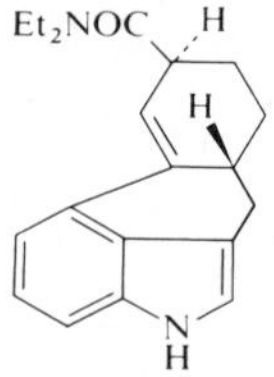   (58)

§ 8.3   Biosynthesis and metabolism of some heterocyclic natural products

Interest in biosynthesis, which commenced with Sir Robert Robinson's observations on the structural relations of natural products, has increased in recent years and many of the hypotheses on how compounds are synthesised in nature have been tested by radioactive feeding techniques. There is a large number of heterocyclic natural products and the biosynthetic pathways to many of the structures have been examined. A very few of these pathways are presented here to illustrate how the natural processes often follow pathways which could be predicted from a knowledge of basic heterocyclic chemistry.

Histidine arises in nature by degradation of a purine as in (8.16) below. The quaternised pyrimidine ring in (59) is obviously susceptible to nucleophilic attack and so hydrolysis can occur, resulting in opening of this ring.

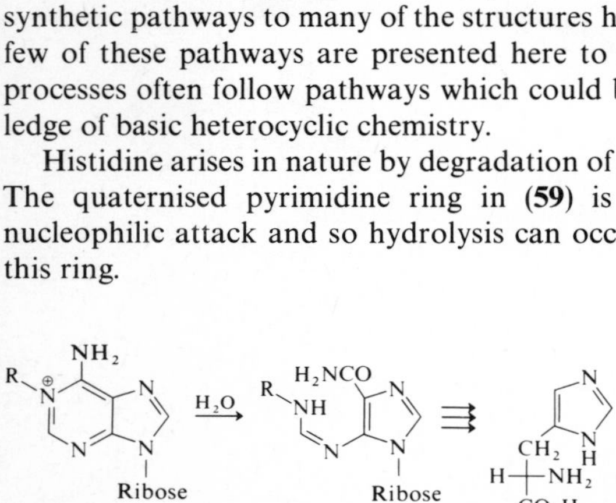

(8.16)

(59)

A variety of naturally occurring heterocyclic systems can result from oxidative cleavage of the indole tryptophan. As we have seen in § 7.2.2, indoles are readily autoxidised in the "pyrrole" ring. The resultant ring-opened compound can give rise to pyridine and quinoline natural products as shown in the sequence (8.17), some of the reactions being similar to synthetic routes to quinolines and pyridines.

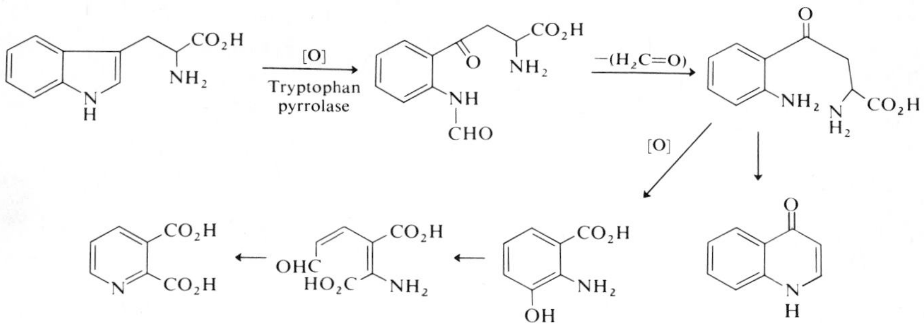

(8.17)

A different type of ring fission of indole has been proposed (cf. (8.18) below) to account for the fact that the quinoline ring in quinine (**49**) is derived from indole precursors. Again the electrophilic reagent attacks at C_3 and this is followed by hydrolysis of the resultant Schiff's base (**60**) → (**61**).

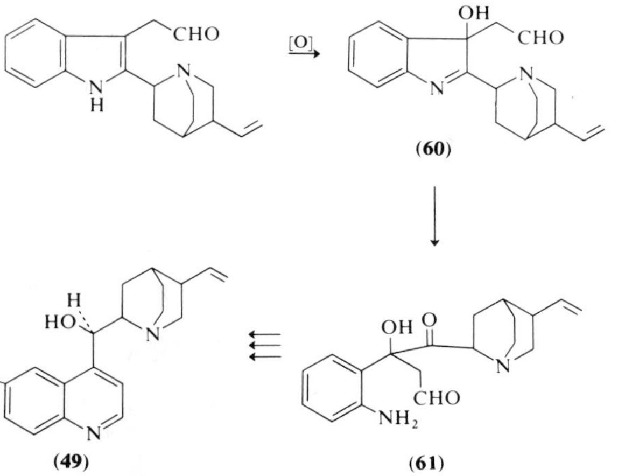

(8.18)

The Pictet–Spengler synthesis of isoquinolines (§ 7.4) is mimicked in the biosynthesis of isoquinoline alkaloids such as papaverine (**56**) (cf. (8.19)). The biosynthesis of a great many indole alkaloids follows a similar pathway to (8.19) in the first step, where the aldehyde secologanin (**62**) forms a Schiff's base which, since C_3 of the resultant indole is occupied, cyclises at C_2 (cf. (8.20) below).

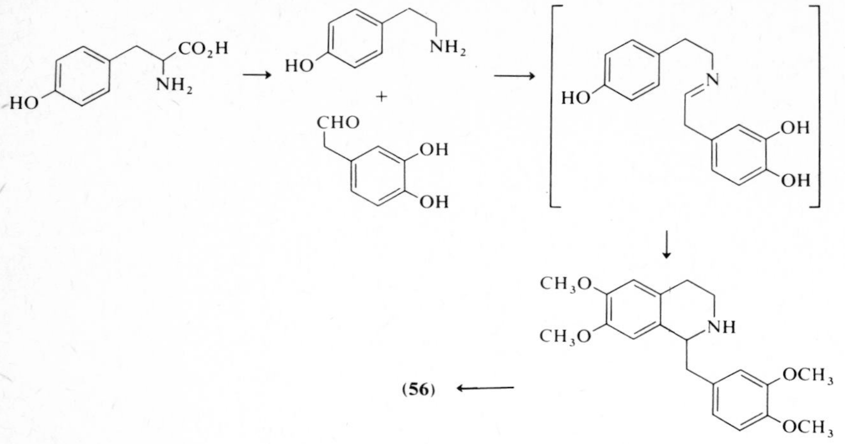

(8.19)

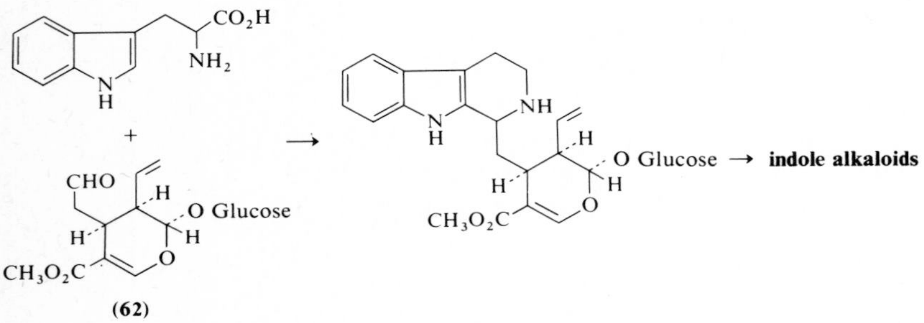

(8.20)

An interesting example of the activation of alkyl groups α- to nitrogen in π-deficient heterocyclic compounds is seen in the biosynthesis of flavin from two moles of 6,7-dimethyl-8-(1'-D-ribityl)-lumazine (63) (cf. (8.21) below). These methyl groups in one molecule react with the equivalent of 2,3-diketobutane from the other molecule to yield the flavin.

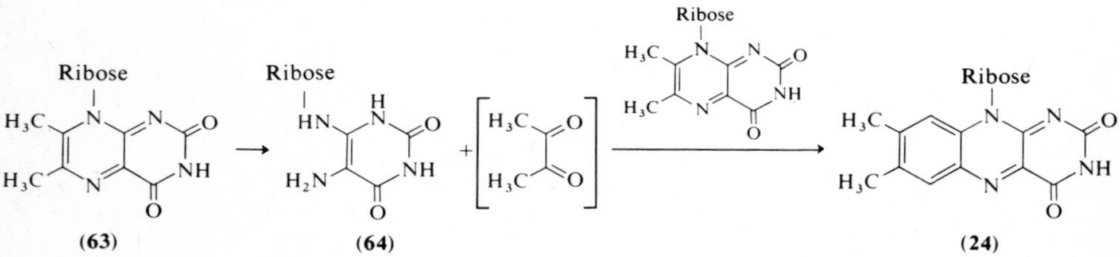

(8.21)

Further Reading

General

H. R. Mahler and E. H. Cordes, *Biological Chemistry*, second edition, Harper and Row, New York, 1971.

Nucleic acids
E. Chargaff and J. N. Davidson, eds. *The Nucleic Acids*, vols. 1–3,
    Academic Press, New York, 1955–1960.

Chymotrypsin, thiamine, NAD and biotin
T. C. Bruice and S. Benkovic, *Bioorganic Mechanisms*, vols. 1–2,
    Benjamin, New York, 1966.

Heterocyclic compounds in medicine
A. Albert, *Selective Toxicity*, 4th edition, Methuen, London, 1968.
W. A. Sexton, *Chemical Constitution and Biological Activity*, 3rd edition,
    E. and F. N. Spon, London, 1963.

β-Lactam antibiotics
E. H. Flynn, ed. *Cephalosporins and Penicillins, Chemistry and Biology*,
    Academic Press, New York, 1972.

Index

Page numbers in **bold type** indicate a fuller discussion of the subject.